APPLICATIONS OF SILICON PHOTONICS IN SENSORS AND WAVEGUIDES

Edited by **Lakshmi Narayana Deepak Kallepalli**

Applications of Silicon Photonics in Sensors and Waveguides
http://dx.doi.org/10.5772/intechopen.71590
Edited by Lakshmi Narayana Deepak Kallepalli

Contributors

Trung-Thanh Le, Salah Rahmouni, Lilia Zighed, DeGui Sun, Didac Vega, Angel Rodriguez, Lakshmi Narayana Deepak Kallepalli

Notice

Statements and opinions expressed in the chapters are these of the individual contributors and not necessarily those of the editors or publisher. No responsibility is accepted for the accuracy of information contained in the published chapters. The publisher assumes no responsibility for any damage or injury to persons or property arising out of the use of any materials, instructions, methods or ideas contained in the book.

First published in London, United Kingdom, 2018 by IntechOpen
IntechOpen is the global imprint of INTECHOPEN LIMITED, registered in England and Wales, registration number: 11086078, The Shard, 25th floor, 32 London Bridge Street
London, SE19SG – United Kingdom
Printed in Croatia

British Library Cataloguing-in-Publication Data
A catalogue record for this book is available from the British Library

Additional hard copies can be obtained from orders@intechopen.com

Applications of Silicon Photonics in Sensors and Waveguides, Edited by Lakshmi Narayana Deepak Kallepalli
p. cm.
Print ISBN 978-1-78984-478-8
Online ISBN 978-1-78984-479-5

Meet the editor

Dr Lakshmi Narayana Deepak Kallepalli is currently working jointly at the National Research Council and the Department of Physics, University of Ottawa; both are located in Ottawa, Canada. After completing his PhD degree on femtosecond laser writing in polymers and crystals from the School of Physics, University of Hyderabad, India in 2011, he moved to the LP3 Laboratory, Marseille, France as a postdoctoral fellow on photonic nanojets project. He moved to Ottawa, Canada in November 2013 and since then has been working in Canada. He has published 23 research articles in international journals, 1 book, and 1 book chapter. His areas of interest include laser-matter interaction, silicon photonics, photonic nanojet lithography, and molecular spectroscopy with expertise in simulations using MATLAB, Python, and knowledge in machine learning.

Contents

Preface

It all started with the famous Moore's law in 1965, which states that the number of transistors on an integrated circuit would double every couple of years. This observation has become a cynosure that continues to drive the electronics industry even today. The density of integrated circuits, popularly known as ICs, has grown tremendously from 1965 to today's high-end microprocessors exceeding 500 million transistors on a silicon chip. In addition to this, the thirst for information and the need to connect globally are spawning a new era of communications. The internet/data era will continue to drive the need for higher bandwidth technologies to keep pace with processor performance. However, computing today is limited by the computer's performance and not by the rate at which data can travel between the processor and the outside world. Here is where the optical communication technology regarding silicon platforms plays a major role.

This book, "Applications of Silicon Photonics in Sensors and Waveguides" (silicon-based photonic technologies), is an essential tool for photonics engineers and young professionals working in the optical network, optical communications, and semiconductor industries. There has been tremendous growth in the research and engineering domains of silicon photonics to address miniaturization and integration of components with diverse functionalities onto a single chip. The revolution first began in the field of electronics especially with transistors and shifted to photonics in the 1980s. We are at the beginning of a new era in communications, one in which silicon photonics may play a significant role.

This book should give readers the foundation on which to participate in this exciting field. When you have read this book, I look forward to seeing you some day on this wonderful journey of "Applications of Silicon Photonics in Sensors and Waveguides". I acknowledge the support from IntechOpen and all the authors who participated in this endeavour.

Dr. Lakshmi Narayana Deepak Kallepalli
Jointly at National Research Council and University of Ottawa
Ottawa, Canada

Chapter 1

Introductory Chapter: Unique Applications of Silicon Photonics

Lakshmi Narayana Deepak Kallepalli

Additional information is available at the end of the chapter

http://dx.doi.org/10.5772/intechopen.78963

1. Introduction

Current technological demands require two key components: miniaturization of devices and integration of multifunctional components onto a single chip offered at low cost. The continuous improvement in meeting the demands of integrated circuits has been enabled by incremental efforts of miniaturization of the transistor [1]. Moore's law states that the minimum feature size shrinks by a factor of 0.7 every 2 years [2, 3]. There has been tremendous growth in the areas of semiconductors and electronics to meet these requirements. However, the research on silicon photonics started only in the 1980s [4]. The advantage of silicon is that its properties can be tailored by doping, which makes it suitable for applications both in electronics and photonics. For useful applications, the technology also plays a major role along with the material. Here, a few applications in photonics domain have been demonstrated.

2. Silicon photonics: a brief overview

Silicon photonics is a disruptive technology, in contrast to conventional technology, as it is vast and has diverse applications. Some important applications include high-performance computing, sensors, and data centers. The photonics industry is rapidly growing to meet the semiconductor and electronics industry. One key advantage can be that of the accessible bandwidth. Most of the electronic devices are limited to GHz speeds in contrast to higher speeds accessible to optical devices. This has spurred researchers develop optical devices operated with faster speeds and at low cost. Silicon photonics is accepted as the next-generation communication systems and data interconnects as it brings the advantages of integration and photonics-high data densities and transmission over longer distances. One potential applica-tion was of waveguides in silicon-on-insulator (SOI) wafer structures in 1985 [5, 6], which was commercialized later in 1989 by Bookham Technology Ltd. [7].

The commercialization for sensor applications began in the 1990s, with integrated gyroscopes and pressure sensors being the first prototype products. Later on, commercialization changed to wavelength-division-multiplexing (WDM) telecommunications products. Here, the low-cost integration capabilities of the platform enabling high-density chips that can perform the multiplexing of many channels of high-speed data onto a single fiber demonstrated the fundamental commercial promise of the technology. The later versions of the data commu-nications advanced the realization of SOI-waveguide p-i-n junction modulators [6] and Ge-, SiGe-based photodetectors, and modulators [8].

3. Role of ultrafast lasers

Ultrafast lasers are known to tailor the properties of materials locally anywhere in 3D to explore salient functionalities. When ultrafast laser pulses (femto and pico) are tightly focused into a material, large peak intensities at the focal volume result in nonlinear absorption and ionization (e.g., multiphoton, tunneling, or avalanche type) guiding to an array of changes in material physical and optical properties. These include negative refractive index (RI) change, positive RI change, or simply void formation. This highly controlled modification endows fs LDW a unique two-dimensional/three-dimensional (2D/3D) microfabrication capability without the use of any phase mask or special sample preparation. Large-scale structures can be fabricated easily by placing the material on a stacked 3D translation stages and control the motion in 3D pattern. In the past, several optical components such as structures for MEMS [9], 2D and 3D gratings [10–17], optical data storage [18–22], waveguides [23–26], photonic band gap materials [27–29], and micro-fluidic structures/devices [30–34]. Ultrafast lasers have been used to change the proper-ties (optical, electrical, chemical, and physical) of silicon toward different applications like surface-enhanced Raman scattering (SERS) for sensors, and waveguides [35, 36]. Silicon also has been tested for its wettability for diverse applications in biophotonics and tissue engineering [36]. There is a need to integrate all these optical components like sensor and waveguides onto Si wafer.

4. Conclusions

The unique combination of properties of silicon combined with photonics technology has been demonstrated in several applications in the past and current. Tailoring the properties of material (silicon) such as bandgap and properties of light such as wavelength, energy, and pulse duration are shown to be the key components in several applications. In this book, some of the key applications in the area of sensors and waveguides have been highlighted.

Author details

Lakshmi Narayana Deepak Kallepalli

Address all correspondence to: lkallepa@uottawa.ca

National Research Council Laboratory for Attosecond Science, Joint University of Ottawa,
Ottawa, Ontario, Canada

References

[1] Shin C. State-of-the-art silicon device miniaturization technology and its challenges. IEICE Electronics Express. 2014;**11**(10):1-11. DOI: 10.1587/elex.11.20142005

[2] Moore GE. Cramming more components onto integrated circuits. Proceedings of the IEEE. 1998;**86**:82. DOI: 10.1109/JPROC.1998.658762

[3] International Technology Roadmap for Semiconductors (ITRS). Available from: http://public.itrs.net

[4] Thomson D, Zilkie A, Bowers JE, Komljenovic T, Reed GT, Vivien L, Marris-Morini D, Cassan E, Virot L, Fédéli J-M. Roadmap on silicon photonics. Journal of Optics. 2016;**18**:073003

[5] Soref RA, Lorenzo JP. Single-crystal silicon: A new material for 1.3 and 1.6 &mgr;m integrated-optical components. Electronics Letters. 1985;**21**:953

[6] Reed G, Headley W, Png C. Silicon photonics: The early years. Proceedings of SPIE. 2005;**5730**:596921

[7] Rickman A. The commercialization of silicon photonics. Nature Photon. 2014;**8**:579-582

[8] Liu J, Cannon DD, Wada K, Ishikawa Y, Jongthammanurak S, Danielson DT, Michel J, Kimerling LC. Tensile strained Ge p-i-n photodetectors on Si platform for C and L band telecommunications. Applied Physics Letters. 2005;**87**:011110

[9] Juodkazis S, Yamasaki K, Marcinkevicius A, Mizeikis V, Matsuo S, Misawa H, Lippert T. Microstructuring of silica and polymethylmethacrylate glasses by femtosecond irradiation for MEMS applications. Materials Research Society Symposium Proceedings. 2002;**687**:B5-B25

[10] Higgins DA, Everett TA, Xie AF, Forman SM, Ito T. High-resolution direct-write multiphoton photolithography in poly(methyl methacrylate) films. Applied Physics Letters. 2006;**88**:184101

[11] Scully PJ, Jones D, Jaroszynski DA. Femtosecond laser irradiation of polymethylmethacrylate for refractive index gratings. Journal of Optics A: Pure and Applied Optics. 2003;**5**:S92-S96

[12] Wochnowski C, Cheng Y, Meteva K, Sugioka K, Midorikawa K, Metev S. Femtosecond-laser induced formation of grating structures in planar polymer substrates. Journal of Optics A: Pure and Applied Optics. 2005;7:493-501

[13] Baum A, Scully PJ, Basanta M, Thomas CLP, Fielden PR, Goddard NJ, Perrie W, Chalker PR. Photochemistry of refractive index structures in poly (methyl methacrylate) by femtosecond laser irradiation. Optics Letters. 2007;32:190-192

[14] Baum A, Scully PJ, Perrie W, Jones D, Issac R, Jaroszynski DA. Pulse-duration dependency of femtosecond laser refractive index modification in poly (methyl methacrylate). Optics Letters. 2008;33:651-653

[15] Hirono S, Kasuya M, Matsuda K, Ozeki Y, Itoh K, Mochizuki H, Watanabe W. Increasing DE by heating phase gratings formed by femtosecond laser irradiation in poly(methyl methacrylate). Applied Physics Letters. 2009;94:241122

[16] Katayama S, Horiike M, Hirao K, Tsutsumi N. Structures induced by irradiation of femtosecond laser pulse in polymeric materials. Journal of Polymer Science: Polymer Physics. 2002;40:537-544

[17] Katayama S, Horiike M, Hirao K, Tsutsumi N. Structure induced by irradiation of femtosecond laser pulse in dyed polymeric materials. Journal of Polymer Science: Polymer Physics. 2002;40:2800-2806

[18] Glezer EN, Milosavljevic M, Huang L, Finlay RJ, Her TH, Callan JP, Mazur E. Three-dimensional optical storage inside transparent materials. Optics Letters. 1996;21:2023-2025

[19] Cumpston BH, Ananthavel SP, Barlow S, Dyer DL, Ehrlich JE, Erskine LL, Heikal AA, Kuebler SM, Lee IYS, McCord-Maughon D, Qin JQ, Rockel H, Rumi M, Wu XL, Marder SR, Perry JW. Two-photon polymerization initiators for three dimensional optical data storage and microfabrication. Nature. 1999;398:51-54

[20] Nie Z, Lee H, Yoo H, Lee Y, Kim Y, Lim K-S, Lee M. Multilayered optical bit memory with a high signal-to-noise ratio in fluorescent polymethylmethacrylate. Applied Physics Letters. 2009;94:111912

[21] Tang H, Jiu H, Jiang B, Cai J, Xing H, Zhang Q, Huang W, Xia A. Three-dimensional optical storage recording by microexplosion in a doped PMMA polymer. Proceedings of SPIE. 2005;5643:258-263

[22] Kallepalli DLN, Alshehri AM, Marquez DT, Andrzejewski L, Scaiano JC, Bhardwaj R. Ultra-high density optical data storage in common transparent plastics. Nature Scientific Reports. 2016;6:26163. DOI: 10.1038/srep26163

[23] Watanabe W, Sowa S, Tamaki T, Itoh K, Nishii J. Three-dimensional waveguides fabricated in poly(methyl methacrylate) by a femtosecond laser. Japanese Journal of Applied Physics. 2006;45:L765-L767

[24] Zoubir A, Lopez C, Richardson M, Richardson K. Femtosecond laser fabrication of tubular waveguides in poly (methyl methacrylate). Optics Letters. 2004;29:1840-1842

[25] Wang K, Klimov D, Kolber Z. Waveguide fabrication in PMMA using a modified cavity femtosecond oscillator. Proceedings of SPIE. 2007;**6766**:67660Q

[26] Ohta K, Kamata M, Obara M, Sawanobori N. Optical waveguide fabrication in new glasses and PMMA with temporally tailored ultrashort laser. Proceedings of SPIE. 2004; **5340**:172

[27] Mendonca CR, Cerami LR, Shih T, Tilghman RW, Baldacchini T, Mazur E. Femtosecond laser waveguide micromachining of PMMA films with azoaromatic chromophores. Optics Express. 2008;**16**:200-206

[28] Zhou G, Ventura MJ, Vanner MR, Gu M. Use of ultrafast-laser-driven microexplosion for fabricating three-dimensional void-based diamond-lattice photonic crystals in a solid polymer material. Optics Letters. 2004;**29**:2240-2242

[29] Straub M, Gu M. Near-infrared photonic crystals with higher-order bandgaps generated by two-photon photopolymerization. Optics Letters. 2002;**27**:1824-1826

[30] Farson DF, Choi HW, Lu C, Lee LJ. Femtosecond laser bulk micromachining of microfluidic channels in poly (methyl methacrylate). Journal of Laser Applications. 2006;**18**: 210-215

[31] Haiducu M, Rahbar M, Foulds IG, Johnstone RW, Sameoto D, Parameswaran M. Deep-UV patterning of commercial grade PMMA for low-cost, large-scale microfluidics. Journal of Micromechanics and Microengineering. 2008;**18**:115029-115035

[32] White YV, Parrish M, Li X, Davis LM, Hofmeister W. Femtosecond micro- and nanomachining of materials for microfluidic applications. Proceedings of SPIE. 2008;**7039**: 70390J

[33] Gómez D, Goenaga I, Lizuain I, Ozaita M. Femtosecond laser ablation for microfluidics. Optical Engineering. 2005;**44**:05110

[34] Day D, Gu M. Microchannel fabrication in PMMA based on localized heating by nanojoule high repetition rate femtosecond pulses. Optics Express. 2005;**13**:5939-5946

[35] Merlen A, Sangar A, Torchio P, Kallepalli LND, Grojo D, Utéza O, Delaporte P. Multiwavelength enhancement of silicon Raman scattering by nanoscale laser surface ablation. Applied Surface Science. 2013;**284**:545-548

[36] Zorba V, Persano L, Pisignano D, Athanassiou A, Stratakis E, Cingolani R, Tzanetakis P, Fotakis C. Making silicon hydrophobic: Wettability control by two-lengthscale simultaneous patterning with femtosecond laser irradiation. Nanotechnology. 2006;**17**(13): 3234-3238

Chapter 2

Macroporous Silicon for Gas Detection

Didac Vega and Ángel Rodríguez

Additional information is available at the end of the chapter

http://dx.doi.org/10.5772/intechopen.76439

Abstract

Macroporous silicon (MPS) has been shown to be a promising material in many areas of technical interest. In particular, MPS has been applied for electronic devices and microfluidic applications. One of the most promising features of MPS is that it enables the development of optical applications using simple and cost-effective technology, compatible with MEMS fabrication processes and suitable for mass production. This chapter describes the application of MPS structures fabricated using electrochemical etching (EE) for the detection of gases of environmental concern in the wavelength range comprising $4 \ \mu m$ to $15 \ \mu m$, such as CO_2. Vertical-modulated MPS structures are reported, whose photonic bandgaps can be placed at different wavelengths depending on the application needs. These structures have been applied to the quantification of CO_2, and these results are summarised here. Detection is performed by the direct measure of absorption, obtaining promising results with short optical paths.

Keywords: macroporous silicon, photonic crystal, electrochemical etching, non-dispersive infrared, gas sensing

1. Introduction

Macroporous silicon (MPS) has been shown to be a versatile material with a broad spectrum of promising applications [1]. MPS was first described by Lehmann in the early 1990s [2–4] and has since attracted great interest among researchers. Of the initial works in MPS development, it is also worth mentioning those from Zhang [5], Propst [6], and Parkhutik [7]. Of particular interest is the *ordered* form of MPS. In such form, pores are etched in a silicon substrate arranged in a uniform pattern. This results in an engineered material whose dielectric constant $\varepsilon(x, y, z)$ has now a characteristic spatial distribution. Given a regular, periodic ε distribution, the propagation of an electromagnetic wave through the media is affected. In particular, with

IntechOpen

an adequately designed pattern, it is possible to create an interference pattern giving rise to *photonic bandgaps* (PBGs) and other optical characteristics for the specific structure. Such structures are thus termed photonic crystals[1] (PCs) [8]. A PBG defines a specific frequency range and, possibly, direction in the crystal, in which an electromagnetic wave propagation is forbidden through the material [9].

Thanks to the existence of PBGs, photonic crystals have been used for advanced applications in optical communications [10], photovoltaics [11], photonics [12], light emission [13], anti-reflection/blackbody emitters [14], and gas sensing [15, 16]. Particularly, in this field, the peculiar functionalities of PCs make these structures very attractive for their use in chemical sensing: gas detection [15, 17, 18] and bio-sensing [19]. The main advantage of using PCs for these areas is their potential to design and fabricate very compact and cheap photonic devices [20, 21]. Furthermore, the possibility of integration into large-scale circuits or microelectronic fabrication processes [21, 22] opens up vast opportunities for novel devices.

In particular, in this chapter, we focus on the use of MPS PCs for the optical detection and quantification of gases. More specifically, the characteristic *absorption lines*, or bands, of a gas in the infrared—that is, the *fingerprint* of the gas—are used to quantify the presence of the gas of interest. MPS demonstrates that is a suitable material and technology to operate in the mid-infrared (MIR), and is useful for the detection of environmental concern gases, such as carbon dioxide (CO_2), methane (CH_4) or nitrogen oxides (NO_x).

2. Sensing application: gas detection

Sensor demand for everyday applications is rapidly growing. The areas of use are many and multidisciplinary [23, 24]—to name a few: environmental [25, 26], safety [27], security [25], health, transportation, and wearables. Market research shows that gas sensor segment has strong growth [24] (forecast to achieve $765 M in 2020 [23]). Furthermore, climate change concerns is making governments and other agencies push the research and deployment of *smart sensor grids* for large area monitoring [28], resulting in new opportunities for emerging technologies such as ultra-compact PC sensors.

2.1. Detection strategies: optical

Optical gas detection provides very desirable advantages over other methods. In first place, pure optical methods like spectroscopy have exceptionally fast response times to changes in mixture concentration. Furthermore, the spectroscopic optical systems are highly selective: they permit identifying a target gas by its *spectral signature* or *fingerprint*[2]. Optical systems can be made extremely sensitive, though this generally requires very long optical paths.

Traditional optical-based measuring is based on the direct measurement of optical power. One of its main drawbacks is that the equipment is large and expensive. Indeed, one has to

---

[1]Nonetheless, MPS is not the only material nor technology in which PCs can be devised and fabricated.
[2]The chemical composition of a gas compound has specific vibrational (atomic bonds; from MIR to VIS) and electronic (electron excitation; from VIS to X-Ray) resonances that result in particular absorption frequencies unique for such gas.

trade-off space for detection limit. Some applications require several centimetres or even metres of optical path to reach the required sensitivity. Furthermore, spectroscopic systems are energy limited, thus they require stable and high power light sources such as lasers or thermal radiators. Spectroscopic systems also require additional complex mechanical and electronic equipment for the signal conditioning and processing. On the other hand, non-spectroscopic optical systems impose less strict requirements on some of these aspects. However, reducing the optical path length has not found a good solution until recently with the advent of PCs.

Other optical systems exist that use alternative detection strategies. Nevertheless, these alternative systems lose some of the more desirable traits of optical detection, like response time and selectivity. Special mention has to be done with respect to *terahertz gas detection*. This technology is closely related to IR optical detection: typical working frequencies range from 0.1 to 10 THz which correspond to wavelengths from 3 mm to 30 μm [29]. At these frequencies, gas molecules have mainly vibrational resonances which are then used in the same manner as for IR optical detection: by absorption measurement. Terahertz technology has been successfully applied for the detection of air pollutants and health applications [30] and long range detection. However, terahertz technology has some serious drawbacks: the principal being the generation of terahertz waves [29].For commercial and industrial applications, *non-dispersive infrared detection* seems to be the most promising optical solution. This is discussed latter in more detail.

2.2. Gases of environmental concern

The growing global consciousness in environmental preservation and climate change has driven the research and development of sensing devices. In particular, monitoring the environment for pollution control [25, 31] is one of the most important applications. Gas sensors are also significant for health [32] and indoor air quality assessment [33].

Air pollutants and greenhouse gases are primarily related to the exhaust gases of combustion processes. Major air pollutants are carbon monoxide (CO), ground-level ozone (O_3), nitrogen dioxide (NO_2), and sulphur dioxide (SO_2). Greenhouse gases are carbon dioxide (CO_2), methane (CH_4), and nitrous oxide (N_2O) [34]. There are many other toxic air pollutants such as nitric oxide (NO), ammonia (NH_3), and hydrocarbons.

These gases have simple molecular composition with strong light absorption in the medium infrared wavelength range. The absorption spectrum of a gas at these light frequencies is caused by the different vibrational and rotational modes of the atomic bonds in the molecules [35]. The absorption coefficient spectra for several of the environmental concern gases are plotted in **Figure 1**. Absorption data of several gases can be found on the freely available HiTRAN database [35]. Each vibrational-rotational mode of a gas corresponds to an absorption line. These modes are narrow and close together[3]. However, at atmospheric pressure *line mixing* occurs, and the instrument resolution further broadens the spectra, transforming the absorption-lines into an *absorption region* or band, as seen in **Figure 1**.

---

[3]An ideal vibration-rotation mode has a single frequency. However, the actual *line profile* of a mode depends on external factors such as the gas pressure, velocity, temperature, etc. Typical line profiles are Gaussian (Doppler broadening), Lorentzian (pressure broadening), and Voigt (mixture of the previous). In standard conditions, ideal gas line profile separation is around $\Delta k = 2$ cm^{-1}, and full-width half-maximum approximately FWHM = 0.2 cm^{-1}.

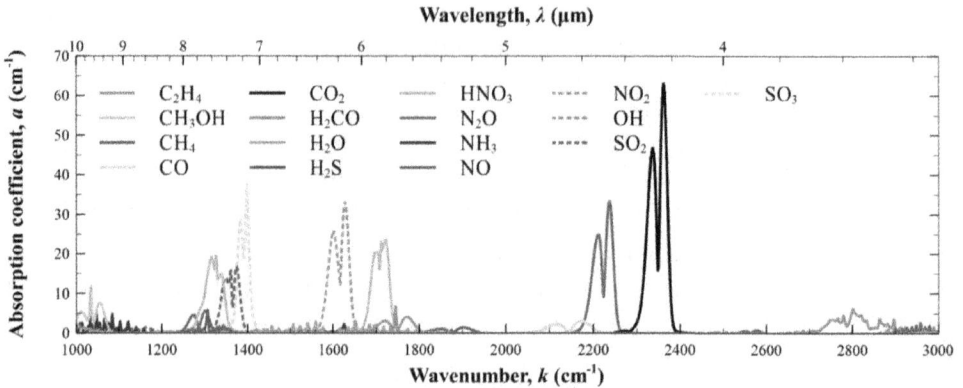

Figure 1. Absorption coefficients of some of the most relevant health and environmental concern gases. The coefficients are calculated from HiTRAN data using an *apparatus function* of $sinc^2(x)$ with an aperture resolution $\Delta f = 4$ cm^{-1}.

From the data shown in **Figure 1**, it is clear that CO_2 is potentially the largest contributor to greenhouse effect. In addition, its absorption band is clearly separated from other strong absorbers, making this gas easy to detect and quantify in an unknown mixture.

2.3. Non-dispersive infrared detection of gases

Non-dispersive infrared (NDIR) detection relies on the fact that certain absorption lines of gases are "isolated" and their wavelengths have little overlap with other gases, as seen in **Figure 1**. This is profited to simplify the design of a gas detector reducing cost, complexity, space requirements, and power [36]. Identification and quantification of a certain gas can be done looking at a narrowband region of the spectrum. This can be done using optical filters or selective light sources, or a combination. At the other end, the detector (a photodiode: PD) will give a measure of the optical power received, directly corresponding to the concentration of the gas. When light passes through a gas mixture, some wavelengths will be absorbed following the Beer–Lambert (B-L) law

$$I(\lambda) = I_0(\lambda)e^{-lca(\lambda)} \tag{1}$$

where I_0 is the optical intensity of the source (or the reference value), I is the received optical power, l is the optical path, c is the gas concentration, and a is the absorption coefficient of the gas; source intensity and absorption coefficient depend on wavelength λ. An optical filter thus selects the absorption lines to measure. This allows for a multigas detection scheme using different filters for each gas [37]. Alternatively, the absorption can be measured as follows: $A = 1 - T = 1 - I/I_0$ (where A is the absorption, and T the transmission). This is valid under the assumption that no light reflection occurs at the gas interfaces[4].

---

[4]The construction of a gas cell as well as other optical elements existing in the light path introduces several reflections. However, all of them are accounted when measuring the reference value I_0.

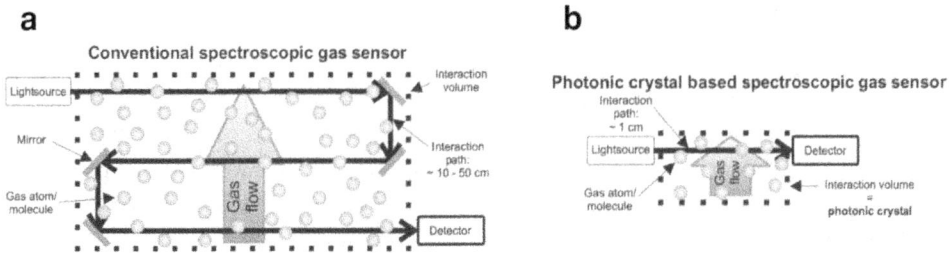

Figure 2. Conventional optical gas sensor using mirrors to attain a long optical path. A PC-based gas sensor can be made much more compact exploiting the special properties of PCs such as slow light. Reprinted from [18], with permission of SPIE.

The simplest NDIR system is depicted in **Figure 2a**. A sample of the unknown gas mixture is placed in a gas cell. Light is then passed through the cell from a source to a detector. From Eq. (1), it is clear that optical path length l is critical for the performance of the system. The necessary optical length l to achieve a desired detection limit can be several meters. Such long paths are highly impractical. Typical commercial products employ special cells (e.g., a Harriot cell) or optical systems using several mirrors to extend the effective optical path keeping a small volume.

To remedy the need of complex, bulky, and fragile optical systems to achieve the long optical paths required, a PC can be used as proposed in **Figure 2b**. PCs-based sensors are projected to require very small footprints, a few centimetres at most. The idea is to take advantage of the special features that PCs exhibit. One of such is the existence of propagation modes with extremely slow group velocity v_g (also called *slow light* modes,) such that interaction time of the propagating electromagnetic wave is increased. Another possibility is the inclusion of defects in the PC to create resonant states. These are localised [38, 39] and can have very high-quality factors (Q-factor) prolonging the interaction time of light with the gas. Other alternative sensing methods can be used where PCs can enhance the system response, e.g., photothermal [40, 41] or refractive index change. On the other hand, PCs can be simply used as highly selective filters.

3. Macroporous silicon applied to gas detection

Macroporous silicon is a *structured* material which has some very desirable properties. In particular, MPS has been extensively applied for the fabrication of PCs. In the first place, silicon is widely used in the industry, and many techniques and tooling exist. Additionally, MPS can benefit of established fabrication processes such as those used for micro electro-mechanical systems (MEMS). For dedicated applications, there are available fabrication technologies suitable for mass production, like electrochemical etching (EE). Given these advantageous traits, MPS has been proposed as a revolutionary material in many application areas. One of such areas is gas sensing. There has been some very intense work in order to develop the technology of MPS as gas sensing devices or as enhancers.

3.1. Seminal works

One of the first uses of MPS for gas sensing is in the work by Geppert [18]. From the analysis of the photonic band structure[5] of the PCs, they observe that some bands are very flat at certain wave vectors, thus the group velocity v_g reduces to values about $0.02c$ to $0.05c$ (c, the speed of light). The slow propagating light has more time to interact with the media making possible to reduce the physical path length while maintaining an acceptable interaction time. It is also observed that some bands have the maximum field intensity located in the high dielectric region (*dielectric bands*), while other bands have the maximum in the low ε region (*air bands*). Thus, air bands must be used for the maximum interaction.

Some of the described devices are three-dimensional (3-d) PCs made of MPS. For these devices, light is shone normal to the surface, propagating parallel to the pores' axis. The devices were used to detect ammonia (NH_3) and sulphur hexafluoride (SF_6). Transmitted light was measured through the PC and a small cavity of 300 μm depth filled with gas. They show promising results, but they argue that the small figures obtained are due to the low coupling efficiency of the PC. Furthermore, no enhancement was observed by the presence of the PC. This is also attributed to the large change in the effective refractive index[6] (n_{eff}) as for high contrast materials the transmittance $T \approx 1/n_{eff}$, thus reducing the coupling efficiency. In [18], it is also discussed a simple way to compensate for deviations in the bandgap due to fabrication tolerances: tilting the sample $15°$, a shift as large as 35 cm^{-1} of the PBG (about 4%) can be achieved.

They also present an alternative structure, refined in later works [17, 42], which uses two-dimensional (2-d) MPS structures. These are easier to fabricate and tolerances are better. As the structure is two-dimensional, the photonic band structure exists only for *in-plane* propagation. Thus, light must be coupled to the *side* of the PC. An example of such arrangement is shown in **Figure 3a**. Additionally, small v_g implies large n_{eff}, diminishing coupling efficiency. To avoid this problem, a special *tapered* section is placed on the interfaces of the PC. This tapered section gradually adapts the air n_{air} to the n_{eff} of the mode. This section can be done by deforming the lattice (elongating one axis) [18] or by leaving additional bulk material at the interfaces (promoting surface waves) [17]. The measurement results of these samples, see **Figure 3b**, show that in this case the PC does enhance moderately the detection of CO_2 (assumed 0 to 100% change). The devices have a 15 dB/mm *insertion loss*, and variability in the enhancement factor is high. These are attributed to uncertainties during fabrication: a fluctuation of 1% of the pore diameter is reported.

3.2. Structure design

From the possible strategies when using MPS structures as gas sensors, 2-d crystals impose the need to inject the light from the sides. In this way is easy to obtain long optical paths along the

---

[5]The band structure is the reciprocal of the dispersion diagram, thus the slope of the band represents the group velocity $\mathbf{v}_g = \partial\omega/\partial\mathbf{k}$, where ω is the wave frequency and $\mathbf{k}$ the wave vector in the reciprocal lattice of the PC, for a plane wave propagating through the media [62].

[6]There exist several methods to find the effective refractive index of a PC. The simplest one for low frequencies is calculating the average of the different materials that compose the PC. To obtain the n_{eff} at higher frequencies, a theoretical study of the photonic band structure is needed, from it, n_{eff} can be derived from the group velocity.

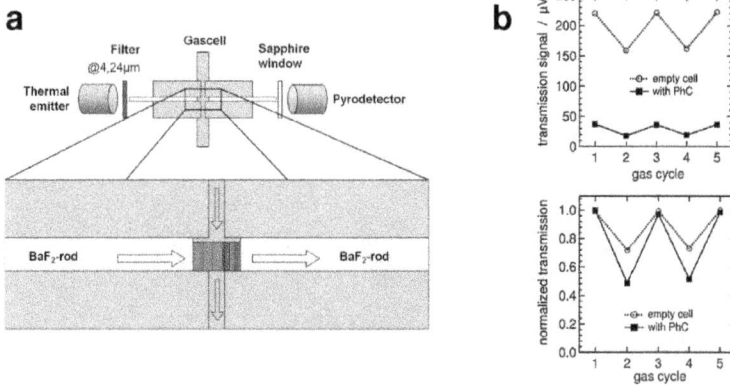

Figure 3. Optical setup for the measurement of gas using 2-d a MPS PC (a). (b) Shows results of measurement of CO_2. The top panel shows the absolute values, while the lower panel shows the relative variation of the measured value. Cycles 2 and 4 correspond to the presence of CO_2, while the other cycles are with reference atmosphere. Reprinted from [17], with the permission of AIP Publishing.

samples, but light coupling is complicated. However, from a practical standpoint, a system based on the normal incidence on the sample, using 3-d PCs, is easier to assemble, align, and calibrate[7]. In this chapter we expose results based in this option with PCs produced with macroporous silicon.

The designed MPS silicon structures are simulated using a simplified model by the finite-difference time-domain (FDTD) numerical method. Some MPS structures will be used as reflectors, while others will be used as filters. Reflection design requires a PBG encompassing just the absorption region of the desired gas. On the other hand, transmission design needs a PGB as wide as possible, with some crystal defects to block all light but the corresponding to the gas absorption line. The largest PBGs are obtained with an opal like PC [43]. As the samples will be illuminated from the top[8], the horizontal pore arrangement is not as important. Our samples pore disposition is a square array. The EE of silicon does not allow getting perfect spherical shapes (opals), but the actual pore profile has a sinusoidal-like shape, or *skewed* sinusoidal shape as seen in **Figure 4**. For the simulations, this profile is approximated by a cylinder and an ellipsoid, or by cylindrical slices, see **Figure 4**.

To place the PBG at the CO_2 absorption line, the vertical period is calculated to be around 700 nm. As a rule of thumb, in a silicon PC, optical bandgaps can be found at about four times the pitch. Therefore, to have a PBG at $\lambda_0 = 4.25$ μm, $l_0 \approx 1.1$ μm. Considering the MPS porosity[9] $\bar{p}$ is about 40%, the effective refractive index $n_{eff} = n_{Si}\bar{p} = 1.48$ [44], where $n_{Si} = 3.47$ [45]. This gives that the vertical period has to be $l_{Si} = l_0/n_{eff} = 718.0$ nm.

---

[7]Indeed, using 3-d PCs, the light can be launched into free space and coupled into the crystal from the top or bottom surface, which are much larger than the sides of the MPS structure.

[8]Concretely, the "top" surface is the surface from where the pores are etched. This surface has the initial pattern of the pore sites. For prime quality wafers, this face is polished and the incident light will have little scattering.

[9]Porosity is defined as the ratio of air to silicon volume in the unit-cell.

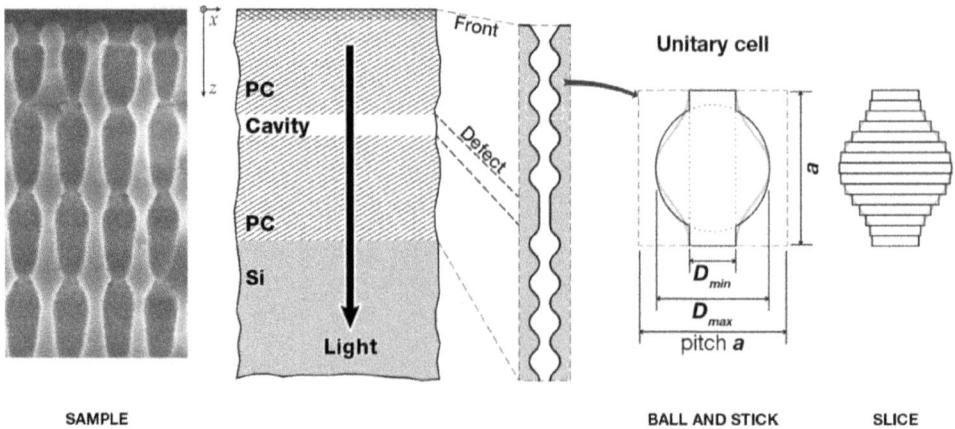

Figure 4. Sample model. The unitary cell can be approximated by a "ball and stick" model consisting of a cylinder and an ellipsoid, or by a slice method, using short cylinders.

The FDTD simulations show that in the vertical direction there is a narrow PBG (7.1%). A *chirped* modulation smoothly varying the vertical period can double the PBG width [46]. For the samples with included defects, initially the photonic response for a single defect was studied. As a further step, the use of several defects has also been analysed. In particular, the coupling between resonant cavities creates additional resonance modes [47]. Using two cavities, the fundamental resonant mode splits into two, making it possible to filter two separate frequencies using a simple PC design, to perform a differential measurement. One of the resonances selects the gas absorption line, while the other is placed in a "safe" wavelength were it will not be influenced by the media. Such configuration would allow ultra-compact filters or selective emitters.

3.3. Fabrication

For MPS fabrication, the EE method is one of the most versatile, and one of the preferred for three-dimensional structure definition. A general description of the method is the dissolution of silicon by an electrically promoted reduction–oxidation chemical process (*redox reaction*). The particular implementation of EE used for the fabrication of *ordered* MPS uses a pretreatment of the silicon substrate to define the sites (*nucleation points*) where localised etching of the material occurs. This method was first described by Lehmann in the early 1990s [3, 4]. Succinctly, the silicon substrate to be processed is placed in a bath of hydrofluoric acid (HF). Silicon is unreactive in normal conditions with HF; however, passing a current activates the redox reaction [48]. As detailed in previous works, the reaction exchanges *holes* at the semiconductor-HF interface. Either n-type or p-type doped silicon substrates can be used; however, for the controlled growth of macro-pores[10], n-type substrates are advantageous, as holes are

---

[10]A porous media is termed microporous for pore diameters less than 5 nm, mesoporous for 50 nm $> d_{pore} >$ 5 nm, and macroporous if pores are larger than 50 nm wide.

minority carriers. To etch the substrate holes must then be generated somehow. The most common way is by using *photogeneration* (light assisted EE). This provides great control on the etching process.

The etching setup places the patterned face to be etched in contact with the electrolyte, opposing to the cathode. The back-face (i.e., the other side) is contacted electrically to close the electrochemical circuit. The illumination can be provided from either the front or back surfaces. For practicality reasons, it is simplest to irradiate from the backside. Our etching setup follows these principles, plus the IR light is provided by IR LEDS with $\lambda_{LED} = 950$ nm, and HF is continuously pumped through the etching cell at a controlled temperature.

The pore growth and morphology depend on several factors of the etching process. The pore front (i.e., the *tip* of the pores) advances at a speed (etching velocity) relatively unaffected by the electrical conditions; however, the etch rate (i.e., the amount of dissolved silicon) is proportional to the electrical current flowing through the electrochemical cell. Thanks to this behaviour, it is possible to use the EE method to fabricate 3-d MPS structures by the modulation of the current and potential during the etching process [49]. More detailed information on the etching process can be found in the works of Lehmann [2], Zhang [5], and Kolasinski [48].

The MPS photonic crystals developed in this work have all been fabricated using the light assisted EE of silicon using n-type substrates. Having pores arranged in an *ordered* fashion generates internal interference patterns for an electromagnetic wave propagating through the porosified media. These interference patterns will depend on the direction of propagation and frequency; hence, the materialisation of a PC. To arrange the pores in a desired pattern, the method by Lehmann [4] is used. This method prepares the wafers before etching using lithography to define the pore sites. Then nucleation points are created by a short silicon anisotropic etching at the sites. This creates small pyramidal pits in the surface, which promote the electrochemical etching at these spots. EE also allows the formation of very large aspect-ratio (AR, ratio of pore diameter to pore length) pores in relatively short time. For example, AR = 1 : 465 has been reported [12].

After the porosification, some samples were post-processed to create a membrane. The membrane is done by anisotropic etching of the back face. For gas measurements, an open membrane is the best option: gas can flow freely through the MPS sample, solving any issues with gas trapping or residence time for the sensors.

3.3.1. Fabrication quality

In general, the EE of silicon produces good quality porous silicon. What is "good quality" with respect to MPS? In a minimal sense, macroporous silicon is of "good quality" if pores are of the same shape and have grown uniformly. Of course this does not provide much information whether the "quality" is acceptable for the intended application. In particular, for photonic applications, the requirements can be very strict. Fabrication imperfections arise due to numerous reasons: wafer crystalline defects, crystal alignment, and local dopant distribution; but also the etching process itself can account for some variability, and lithography errors will also cause flaws in the grown pores. The common fabrication defects one can encounter in a MPS structure fabricated by EE are summarised in **Table 1**.

Large optical influence		Little optical influence	
![]	Pore death	![]	Spiking
![]	Pore branching	![]	Unstable growth
![]	Axis alignment	![]	Microporous layer
![]	Cross-section		
![]	Profile fidelity		

Table 1. Defects found in macroporous silicon.

From **Table 1**, the defects that are of greater concern are the ones classified as "large optical influence." Here *surface roughness* is not considering as a defect. It is a consequence of minute variations inherent in any fabrication process. The amount of perturbation does not alter the pore morphology or cause long-range effects and will not give unexpected results.

Pore death is the premature ending of a pore resulting in a shortened pore with respect to its neighbours. This can be single, isolated pores or can affect several pores. *Current starvation* is the most common cause. During etching, there is a minimal current under which there are not enough carriers to sustain stable growth. In this situation, current will concentrate on some pores resulting in wider, distorted, and shorter pores. The solution is increasing the etching current. Also, shorter structures are less susceptible to suffer from pore death.

Pore branching is generally a consequence of pore death but can also happen if the etching parameters are not adequate: if applied potential is too small, the space charge region (SCR) will be small and unintended holes could find their way to spots in the pore wall far from the pore tip and secondary pores grow from that points.

Pore axis alignment depends on the crystalline direction of the Si substrate. Pores preferentially develop following the principal crystalline directions of the silicon substrate; however, the pores also have a tendency to grow following the current paths. If the crystal is severely misaligned, the mismatch between the crystal direction and the current direction will cause that the pores grow in several directions simultaneously [50]. The pore *cross-section* affects the symmetry of the PC. In the stable growth phase, cross-section approximates a circle; however, under certain condition of applied potential and electrolyte additives, the cross section can change its shape.

Spiking is the appearance of random streaks of nanometric branches, spreading out in random directions and with irregular shape. Excessive voltage during the etching promotes spiking. Moderate to low spiking has little influence on the optical performance. However, it changes the porosity, which may shift the wavelengths of the optical features. *Unstable growth* is basically caused by substrate inhomogeneity, though the dynamic behaviour of the etching system may also influence [51]. In general, unstable growth is mild and can be observed with pores not evolving in a straight line but slightly wandering. Another form of unstable growth is the *long-range* variation of pore diameter.

Profile fidelity refers to the closeness of the actual pore profile to the intended profile. It is considered a defect because it is an unintended result which affects *severely* the optical response of the etched sample. Several test runs with progressive refinements are needed before the near-optimal conditions and waveforms are found. After EE, the pore's walls may be covered with a layer of *microporous silicon*. This layer will depend on the electrolyte concentration and present additives. It has no noticeable optical influence in the MIR, and it can be left; however, it is very easy to remove by a quick dip in HF.

The abovementioned discussion was made considering the local effect of perturbations in the pore shape and outcome of the etching process. Nonetheless, the uncertainties of etching, the inhomogeneity of the substrate, and the flaws in lithography can be considered on a larger scale or *long-range*. Pores at distant locations[11] are also affected by the etching system. The most common long-range effect is the variation of pore diameter of up to a few percent [42]. Also, the etching process generates H_2 bubbles which must be removed otherwise they adhere to the surface and create *big* spots where the etching stops. More in-depth analysis can be found in the works by Föll and co-workers [52, 53].

3.4. Absorption and losses of macroporous silicon

One important concern is determining how absorption and fabrication tolerances will affect the performance of the detection system. Intrinsic silicon absorbs light at wavelengths shorter than 950 nm, approximately, however when silicon is doped with impurities, strong absorption may be observed in the MIR range. Several papers report on the absorption dependence with doping level in silicon [54]. Interestingly, for the doping concentrations used in the fabrication of optical gas sensors working in the middle infrared ($n < 5 \cdot 10^{17}$ cm^{-3}), it is found that material absorption is negligible [54, 55].

On the other hand, fabrication tolerances definitely do have an impact on optical performance. For instance, in [17], it is reported that a 1% variation of the pore radius attenuates 15 dB/cm the transmitted light. This is further studied in [42], global pore diameter fluctuation of 3.5% was caused by the spatial variation of dopant. They further claim that, for a better than 90% transmission in a 1 mm thick PC, the pore position should change less than 0.3% and diameter has to be within 0.5%. A systematic study of the effect of in-plane disturbances in electrochemically etched MPS is presented in [56]. Some interesting conclusions extracted from this work are that ellipsoid perturbation up to $\pm 10\%$ has little impact on the PBG if the porosity is maintained near its optimum, and that a small variation of the air fraction of just $\pm 4\%$ is enough to shift a narrow PBG ($\Delta f = 5\%$) out of the designed central gap frequency f_0. Similar conclusion was found in the work by our group [57], where the effect of vertical period variation in 3-d MPS was analysed. Wide PBGs were designed for the samples used in this work, but noticeable narrowing and shift could be observed with periodicity variations above

---

[11]For example, a scratch on the back-surface or shadow of the illumination will cause the shape of such scratch or shadow to be transferred to the pores grown at the front surface: the affected area will have smaller diameter pores or, in extreme cases, dead pores and branching.

±5%. Furthermore, it was found that the PBG became more transparent allowing light transmission higher than a 10% in the forbidden band.

4. Measurements and results

Samples of MPs photonic crystals were fabricated with a lattice pitch of 700 nm and modulated pore profile. The fabrication conditions were the ones described earlier in Section 3.3—using the EE method. The modulation profiles for the 3-d PCs were programmed to generate a "strong"[12] profile having a 700 nm vertical period, the same as the lattice pitch. The samples were initially characterised to obtain their optical spectrum in the MIR range using a Vertex 70 Fourier-Transform Infrared Spectrometer (FT-IR) from Bruker Optics. Also some of the structures were cleaved and latter inspected by SEM to determine the pore morphology and actual etched profile. After the samples were analysed, the valid samples were used in a dedicated gas measuring system also using FT-IR spectrometers to measure the performance of the PCs in sensing carbon dioxide.

4.1. Fabrication, sample characterisation, and morphological study

The fabricated samples were etched using a silicon substrate with ⟨001⟩ crystal orientation. For the 700 nm period MPS structures, substrates of the appropriate resistivity were used. The pores are arranged in a square pattern. This pattern is transferred to the substrate surface using nano-imprint lithography (NIL). The etching temperature was set to $10°C$. The resulting PCs are to be used with light coming from the top, aligned to the pores' axis, with normal or quasi-normal angle of incidence (θ_i from $0°$ to $13°$). The actual modulation waveforms depend whether the samples will be used for reflectance or transmittance. For the samples used in transmission gas measurements, membranes were made. However, the structure resilience is critical given the thin porous layer, so membranes are *not* open. Indeed, the porosified depth of the MPS crystals ranges from 20 to 50 μm. Such thin layers are very fragile. For this reason, membranes did not reach the pore tips.

4.1.1. Morphological analysis

The morphological analysis of some samples shows, as seen in **Figure 5**, that the etched profiles of the MPS structures have some imperfections. The observed flaws are mainly small variations in pore radius between adjacent pores, pore "wiggling", skew, and pore death. On the surface of the pores, a rough finish due to microporous silicon can be appreciated (see **Figure 5**). Pore radius variation is small—less than 10%—for the samples used here. This has a small impact on the optical response of the PC as has demonstrated by our works [57] and others [58, 59]. Skew is generally not an issue. Some extreme cases have shown up to 5° of crystal misalignment from the [001] direction (see **Figure 5b**), but otherwise pores grow fine. Some pore "wiggling" can be observed in the fabricated samples (see **Figure 5a**). As a

---

[12]This profile tries to obtain a spherical shape as close as possible.

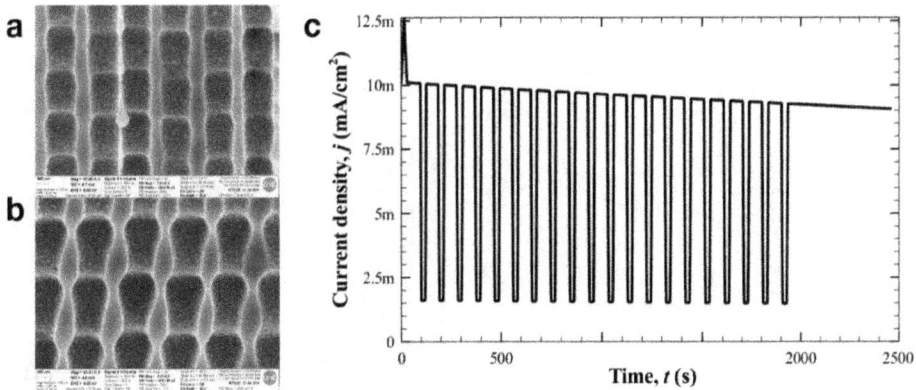

Figure 5. Cross section detail of a MPS photonic crystal (a) and (b), and the current waveform used to generate (a).

consequence, there is some decrease of the optical performance of the PC. In general, the degree of wiggling in our samples is small.

The fabrication process for the MPS samples has been optimised to obtain structures virtually free from dead pores. In general, optical performance is preserved if dead pores are few, or if pores die late during etching. To avoid pore death in the initial samples, the lower current limit was increased, so dynamic range was reduced and modulation could not be as strong as desired (compare **Figure 5a** and **b**).

The most significant defect found in the samples here used was vertical period variation[13]. As seen from **Figure 5**, for neighbouring pores, the beginning and ending of each vertical period is slightly different. As shown in our previous work [57], this variation has a noticeable influence on the PC optical response. In general, the 700 nm samples fabricated with our equipment show about 5% variation in periodicity which results in about 30% PBG narrowing from the ideal and nearly 10% minimum transmittance [57][14].

Two types of PC structures were fabricated for this work. Samples used for reflectance measurements are regular 3-d structures with a continuous sinusoidal-like profile for the pores. Such samples are shown in **Figure 5** along with an example of current waveform used to generate them. The cross section images reveal that the modulation is slightly skewed to the beginning of the modulation. The vertical period is measured to be 700 nm which is the desired length. Total etched depth is 25 μm and the mean pore diameter is $\bar{d} = 500$ μm. The programmed "modulation index" [15] is about $m_{in} = 0.7$, but results in $m_{etch} \approx 0.1$. Despite this, this structure has good enough optical performance.

---

[13]That is for the current way the PCs are being used: with light coming from the "top" along the pores' axis.

[14]These performance figures were achieved with the latter fabricated samples, with an optimised EE process.

[15]The input waveform is a square signal, so the concept of modulated index in AM is extended here as $m = \max[r(t)]/\min[r(t)]$, where $r(t)$ is the radius of the poresuperfluous.

a

b

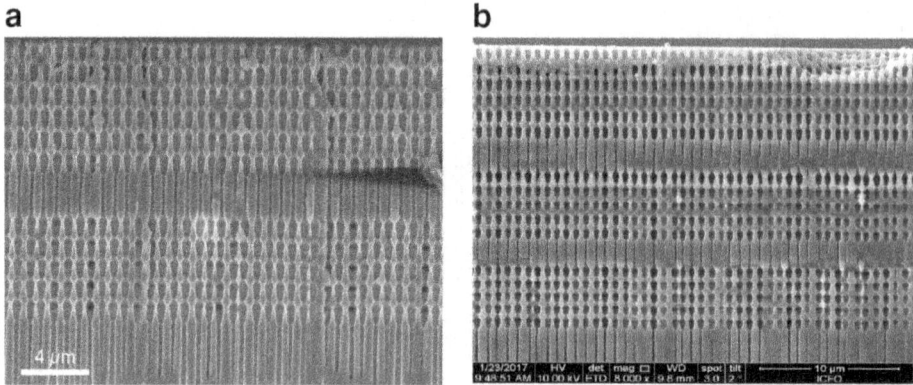

Figure 6. Macroporous silicon structures used in transmission measurements. (a) Is a single defect PC, while (b) shows a two defect MPS structure. Panel (b) © 2017 IEEE. Reprinted, with permission, from [60].

Comparing the etched pore profile from the input waveform (**Figure 5c**), it can be seen that during the high plateau portion of the profile, once the pore has reached its maximum diameter, it slowly starts narrowing as the pore front advances. Presently, this can only be "corrected" by trial and error and judicious changes to the current and potential waveforms. For example, a second profile was designed with smoother transitions and "pre-skewing" resulting however in the PC of **Figure 5b**. Better modulation index was obtained ($m_{etch} \approx 0.2$) but the skewing did not improve. In addition, the vertical period increased to 850 nm, which corresponds to a PBG centred at $\lambda_0 = 5.03$ μm.

Samples used for transmittance measurements include one or more defects, as shown in **Figure 6**. These structures have been fabricated using a refined profile. As seen on the SEM cross-section micrograph, the modulation skew is still present[16], but the vertical period length is better adjusted. Otherwise, the quality of the fabricated structures is maintained. The modulation index has been increased, with pores having a smaller diameter at the necks. This greatly improves the PBG (both width and transmittance blocking) of the PC compared to the first structures depicted in **Figure 5**. The vertical period of the PC used for these samples is $\bar{p}_v \approx 1$ μm. These samples have at least one planar defect in the regular PC profile to define a resonant cavity. These cavities have a diameter $d_{def} = 230$ nm, and length was varied around $l_{def} = 2.1$ μm [60], to place the resonance at $\lambda = 4.25$ μm.

4.1.2. Optical response

Samples were optically characterised after fabrication to ensure the adequacy to sensing CO_2. This was done in a FT-IR spectrometer in reflection mode with light incident at a quasi-normal angle of $\theta_i = 13°$. Also some samples were analysed in transmission mode with light illuminating in normal angle of incidence. The wavelength range of the analysis extends the MIR and some of the NIR regions from $\lambda = 1$ μm to 20 μm.

---

[16]However for this particular instance, the asymmetric modulation was actually designed.

Samples used for reflectance gas measurements, such as the one in **Figure 5a**, have a single PBG in the optical response. The morphological analysis shows that, as the modulation index is small, the PBG will be relatively narrow. This is confirmed in the measurement shown in **Figure 7**. The results confirm that as expected by the criteria given above, the PBG of a 3-d PC with a vertical period of 700 nm is centred at $\lambda_0 = 4.25\ \mu$m. PC quality and modulation index limit the reflectance to about 30%. In spite of these shortcomings, the obtained response is good enough to be used for the sensing of carbon dioxide.

The samples for transmission measurements have different responses according to the number of PC defects placed. As seen in **Figure 8a, b**, placing two defects gives rise to two resonant

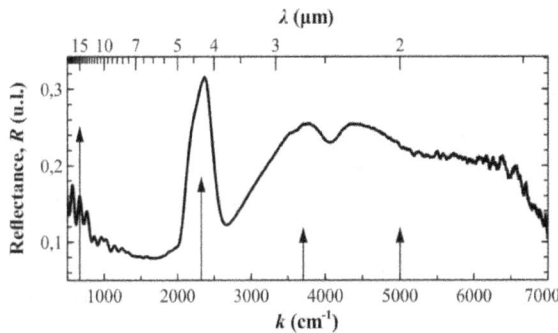

Figure 7. Measured spectra of the PC shown in **Figure 5a**. The absorption lines of CO_2 are marked with arrows, the length giving a qualitative impression of their strength. It can be clearly seen how the PBG of the MPS structure corresponds to the $\lambda = 4.25\ \mu$m absorption line.

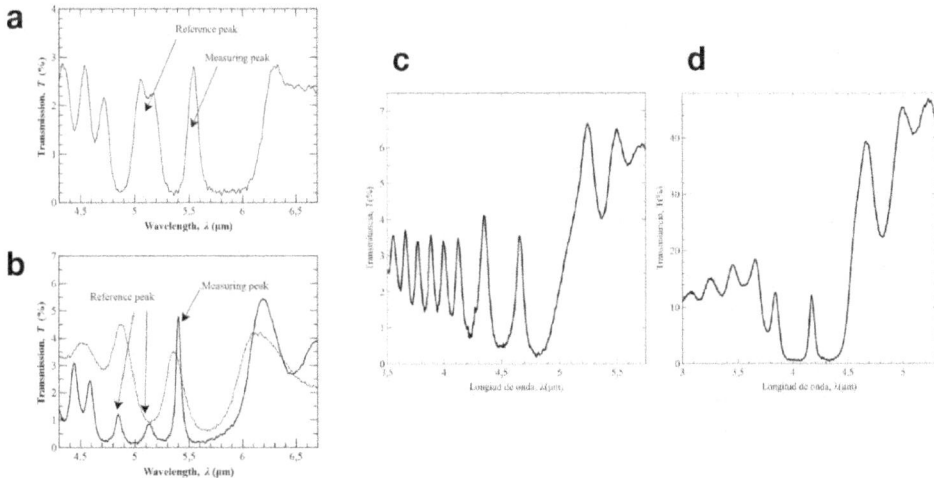

Figure 8. Characterisation of transmission PC samples. (a) Is a two defect sample which shows two resonance states inside the PBG. (b) Is for a three defect PC (thick line) compared to a single defect crystal. (c) Is the characterisation of the sample in **Figure 6b**. (d) Shows the response of a two-defect membrane PC with a peak centred in the CO_2 absorption line at 4.25 μm, and a reference peak at 3.8 μm. Panels (a) and (b) © 2017 IEEE. Reprinted, with permission, from [60]. Panels (c) and (d) from [61] are licensed under CC BY-NC-ND 3.0.

states, and three states if three defects are used. Coupling between resonant cavities induce the appearance of several resonances in the optical spectra even if all cavities are of the same dimensions. The placement of the defect along the PC depth also influences the coupling efficiency and quality factor of the resonance. The creation of a membrane also improves the transmitted signal, compare **Figure 8c** and **d** where the transmitted peak is almost 10% to the 4% of the non-membrane sample.

4.2. Gas measurement system

After the devices were characterised, they were measured under different atmospheres consisting of CO_2 diluted in pure nitrogen. The amount of carbon dioxide was controlled using a two gas mixer in which one line is pure nitrogen as the carrier gas, and the process gas (CO_2 is this case) is connected to the second line. Each line is controllable separately: shutting, pressure, and flow. Both gas lines are then mixed in a manifold and the mixture is then output to the gas cell. The gas cell is sealed with one input port and one exhaust port. The flow control is performed using mass flow controllers (MFCs) of different range: the carrier gas controller (MFC1) is 2 l/min full scale (FS), and the process gas controller (MFC2) is 200 ml/min FS. Both MFCs are calibrated for N_2, so for the process gas a correction factor is used. For the measurements, a constant flow rate of $\dot{V} = 400$ ml/min for the mixture is used. In these conditions, the minimum gas concentration goes from 7500 ppm to 50%. Given the performance data of the MFCs, the concentration uncertainty is given by $B_c^2 \approx c^2 \left[3.15 \cdot 10^{-3} + 0.25 \cdot 10^{-3} \left(c^{-1} - 1 \right)^2 \right]$, where $c = \dot{V}_2 / (\dot{V}_1 + \dot{V}_2)$ is the gas concentration, and $\dot{V}_1$ and $\dot{V}_2$ are the flow rates of carrier and gas, respectively. Note that $\dot{V} = \dot{V}_1 + \dot{V}_2$.

The gas mixture is fed to a gas cell purposely built where the MPS photonic crystal is placed. The gas flows continuously through the cell. For the measurements, a broadband infrared light source is used. This light is directed to the cell and then collected by a photodetector (PD). In the work here presented, the MPS structures were characterised by FT-IR spectrometry. The expected response of an autonomous NDIR measuring device can be then extrapolated from the spectroscopic gathered data in the characterisation of the PCs.

4.2.1. Principle of detection and method

The basic idea here proposed for an NDIR system is to use a MPS photonic crystal with a PBG wide enough to comprise one of the gas absorption *regions*. In the case of CO_2, this region extends from $\lambda \approx 4.20$ to 4.35 μm (see Section 2 – above). Ideally, all light outside the PBG should be discarded and not arrive at the detector. Once the light has been filtered, the photodetector gives a reading proportional to the incoming optical power, and further filtered by the detector's photo-response. In the following discussion, it is assumed that the PD response is flat in the absorption region of the gas.

From the characterisation measurements, it is straightforward to approximate the expected response of the NDIR system. The spectra can be any of the reflectance, transmittance or absorbance, as these are proportional to the power of the electromagnetic wave: $|E_r(\lambda)|^2 = R(\lambda)|E_i(\lambda)|^2$, and so on. Now it is possible to define the quantity $S_R = \int R(\lambda) d\lambda$ as the

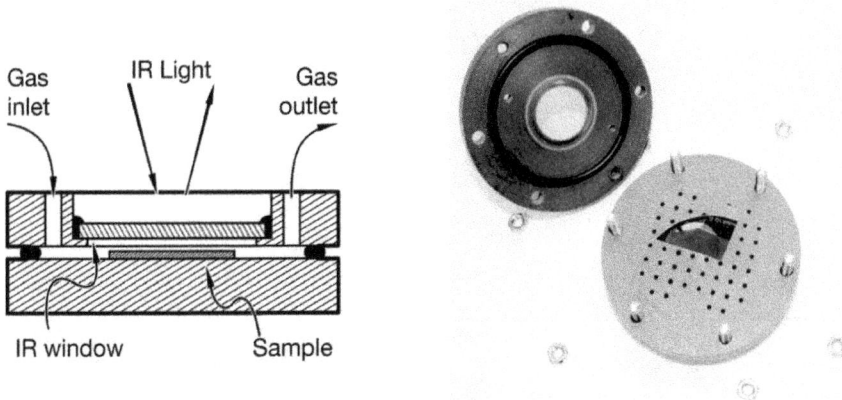

Figure 9. Gas cell for reflection measures. A schematic section view is shown and the actual fabricated cell is pictured next to it. The sample is held in place by locking it with a spring screwed in a suitable position of the cell base grid. The cell is made airtight by an O-ring, and sealant around the IR window.

normalised optical power received at the PD from the reflectance data. S_R is thus proportional to the gas concentration and estimates the actual output of a PD. Equivalent power quantities S_T and S_A can be defined for the transmittance and absorbance spectra respectively. A second measuring criteria is simply evaluating the reflectance spectrum at a given wavelength $R(\lambda_{gas}) = R_{gas}$ (equivalent quantities are defined for transmittance and absorbance). Such wavelength is chosen to be where absorption is the greatest. The relation of this quantity with an actual output of a PD is less clear but serves to evaluate the performance of the gas cell and PC.

4.3. Reflection CO_2 measurement

Reflection measurement of CO_2 uses a PC as a selective reflector. The reflection spectra are obtained from which the absorbance spectra are calculated as $A(\lambda) = 1 - T(\lambda) = 1 - R(\lambda)/R_0(\lambda)$; where $R(\lambda)$ is the measured spectrum at some gas concentration, and $R_0(\lambda)$ is the reference spectrum (only N_2). The "optical power" is then $S = \int A(\lambda)d\lambda$. Using a reference spectrum removes any effect of the optical system, such as reflectance at optical interfaces.

4.3.1. Setup

The measurement setup consists of the gas mixer connected to a specifically built gas cell. A schematic view of the cell design is shown in **Figure 9** alongside the actual built device. The cell consists of two plates separated with an O-ring such that the cell is made airtight. The top plate has one port for gas input and another for output, and a central opening where a potassium bromide (KBr) window is placed and held in place by a sealant. When assembled, the two plates are held by several screw ensuring mechanical stability and airtightness, with the PC underneath the IR window. The gap from the sample to the window is about 0.5 mm, resulting in a total[17] light path of 1 mm.

---

[17]This figure may change due to uncertainty. For example MPS sample thickness or tightening of the screws can change the gap as much as 0.2 mm. It must be remarked that this is a proof-of concept cell.

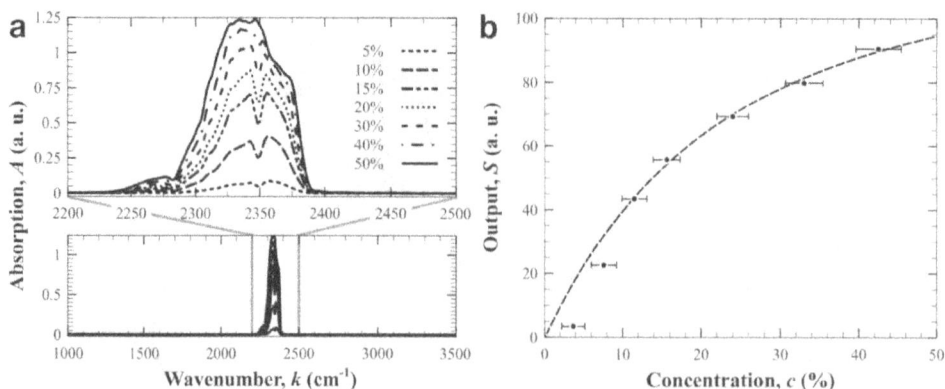

Figure 10. Measurement results for reflectance setup. The spectral response (a) from $k = 1000$ cm^{-1} (10 μm) to 3500 cm^{-1} (2.86 μm) has been normalised to a pure N_2 atmosphere and shows the effect of absorption due to CO_2 in the $\lambda = 4.25$ μm band. The output response (b) against concentration shows good sensitivity for concentrations above 5%. The dashed line is the ideal Beer–Lambert response.

The cell is then placed in the spectrometer to make the gas concentration readings. In particular, a Renishaw Raman microscope equipped with Smiths IlluminatIR FT-IR module is used to take reflectance measurements at normal incidence, $\theta_i = 0°$. The aperture sizes used in the optical setup was 50 μm × 50 μm. The wavelength range extends from $k = 1000$ cm^{-1} (10 μm) to 3500 cm^{-1} (2.86 μm). Resolution is 4 cm^{-1}.Each spectrum at a given concentration was averaged over 256 scans.

4.3.2. Results

Reflectance spectra were measured for CO_2 at concentrations from 3 to 43%, the results summarised in **Figure 10**. Lower concentrations were tested but for concentrations below 3% changes are very small and difficult to discriminate. This effect is then observed in the "optical power" signal S at the output (**Figure 10b**) calculated as mentioned above. **Figure 10b** also shows that the Beer–Lambert law is closely followed by the measured data for concentrations above 10%. From the fitted B-L model, it is possible to extract the exponential term $l \cdot a \approx 3.65$ (where l is the optical path length, and a the absorption coefficient). From the HiTran data for carbon dioxide, at $\lambda = 4.25$ μm and 4 cm^{-1} instrument resolution, $a_{CO_2} \approx 50$ cm^{-1}. This gives that the effective path length is $l_{eff} \approx 0.75$ mm. Measurement uncertainty is dominated by the mixer uncertainty, as seen in **Figure 10b**, as power measures are averaged over a large number of scans. This results in a fairly constant uncertainty across the full measurement range about ±2%.

5. Conclusion

Macroporous silicon is a versatile material that has shown to be a good candidate for the obtainment of PCs for gas sensing applications. Using fabrication methods such as EE opens up the possibility of obtaining high-quality photonic filters, in large quantities, and economically competitive. Furthermore, this fabrication technique is very flexible allowing creating customised designs with little effort. Using silicon as the base material has other benefits, such as the reutilisation of the existing manufacturing tooling and the reuse of process flows. Moreover, EE is compatible with microfabrication technology and might be incorporated in VLSI designs to build complete sensing devices. This will result in more compact and integrated system design lowering the bill of materials, costs, and improving manufacturability.

Here we demonstrate one possible way to use MPS PCs for gas sensing: as selective filters. Carbon dioxide has been detected and quantified using NDIR reflectance measurements. It has been found the MPS crystal has also an effect in the measured absorption. This is due to the very nature of the PC, slowing the group velocity of the incident light and enhancing the interaction time—increasing the effective optical path length—with the gas mixture. The inclusion of resonant cavities further enhances light absorption by inducing resonant states and spatially confining the IR radiation.

The PCs here shown, prove that a compact CO_2 sensor using MPS technology is feasible, achieving a detection range near to that of commercial optical devices based on IR PD/LED. The devices shown here have room for improvement in particular regarding the fabrication, and progress is steadily being made in this area.

Acknowledgements

This work has been funded by the Spanish *Ministerio de Economía y Competitividad* with research grants TEC2010-18222 and TEC2013-48147-C6-2-R. The authors would also like to acknowledge the contribution and help from Daniel Segura García and David Cardador Maza in the preparation of this chapter.

Conflict of interest

The authors declare that there is no conflict of interest.

Author details

Didac Vega* and Ángel Rodríguez

*Address all correspondence to: didac.vega@upc.edu

Electronic Engineering Department (EEL), Universitat Politècnica de Catalunya (UPC), Barcelona, Spain

References

[1] Vega Bru D, Rodríguez Martínez Á. Macroporous silicon: Technology and applications. En: Igorevich Talanin V, editor. New Res. Silicon - Struct. Prop. Technol., InTech; 2017. DOI:10.5772/67698

[2] Lehmann V, Föll H. Formation mechanism and properties of electrochemically etched trenches in n-type silicon. Journal of the Electrochemical Society. 1990;**137**:653-659. DOI: 10.1149/1.2086525

[3] Lehmann V. The physics of macropore formation in low doped n-type silicon. Journal of the Electrochemical Society. 1993;**140**:2836. DOI: 10.1149/1.2220919

[4] Lehmann V. The physics of macroporous silicon formation. Thin Solid Films. 1995;**255**:1-4. DOI: 10.1016/0040-6090(94)05620-S

[5] Zhang XG. Mechanism of pore formation on n-type silicon. Journal of the Electrochemical Society. 1991;**138**:3750. DOI: 10.1149/1.2085494

[6] Propst EK. The electrochemical oxidation of silicon and formation of porous silicon in acetonitrile. Journal of the Electrochemical Society. 1994;**141**:1006. DOI: 10.1149/1.2054832

[7] Parkhutik V. Porous silicon—Mechanisms of growth and applications. Solid State Electronics. 1999;**43**:1121-1141. DOI: 10.1016/S0038-1101(99)00036-2

[8] Ho KM, Chan CT, Soukoulis CM. Existence of a photonic gap in periodic dielectric structures. Physical Review Letters. 1990;**65**:3152-3155. DOI: 10.1103/PhysRevLett.65.3152

[9] Pendry JB. Photonic band structures. Journal of Modern Optics. 1994;**41**:209-229. DOI: 10.1080/09500349414550281

[10] Taalbi A, Bassou G, Youcef MM. New design of channel drop filters based on photonic crystal ring resonators. Optik- International Journal of Light and Electron Optics. 2013; **124**:824-827. DOI: 10.1016/j.ijleo.2012.01.045

[11] Lenert A, Rinnerbauer V, Bierman DM, Nam Y, Celanovic I, Soljacic M, et al. 2D Photonic-crystals for high spectral conversion efficiency in solar thermophotovoltaics. Proc. IEEE Int. Conf. Micro Electro Mech. Syst., 2014, p. 576-579. DOI: 10.1109/MEMSYS.2014.6765706

[12] Schilling J, Birner A, Müller F, Wehrspohn RB, Hillebrand R, Gösele U, et al. Optical characterisation of 2D macroporous silicon photonic crystals with bandgaps around 3.5 and 1.3 μm. Optics Materials (Amst). 2001;**17**:7-10. DOI: 10.1016/S0925-3467(01)00012-X

[13] Cheylan S, Trifonov T, Rodriguez A, Marsal LF, Pallares J, Alcubilla R, et al. Visible light emission from macroporous Si. Optics Materials (Amst). 2006;**29**:262-267. DOI: 10.1016/j.optmat.2005.06.021

[14] Vega Bru D, Todorov Trifonov T, Hernández Díaz D, Rodríguez Martínez Á, Alcubilla González R. Blackbody behaviour from spiked macroporous silicon photonic crystals. Spanish Conf. Electron Devices. «8th Spanish Conf. Electron Devices», 2011, p. 257-258

[15] Vega D, Marti F, Rodriguez A, Trifonov T. Macroporous silicon for spectroscopic CO_2 detection. IEEE SENSORS 2014 Proc., Valencia: IEEE; 2014, p. 1061-1064. DOI: 10.1109/ICSENS.2014.6985187

[16] Cardador D, Vega D, Segura D, Trifonov T, Rodr?guez A. Enhanced geometries of macroporous silicon photonic crystals for optical gas sensing applications. Photonics and Nanostructures – Fundamentals and Applications 2017;**25**:46-51. DOI: 10.1016/j.photonics. 2017.04.005

[17] Pergande D, Geppert TM, Rhein A von, Schweizer SL, Wehrspohn RB, Moretton S, et al. Miniature infrared gas sensors using photonic crystals. Journal of Applied Physics 2011; **109**:83117. DOI: 10.1063/1.3575176

[18] Geppert TM, Schweizer SL, Schilling J, Jamois C, Rhein AV, Pergande D, et al. Photonic crystal gas sensors. En: Fauchet PM, Braun PV, editores. Opt. Sci. Technol. SPIE 49th Annu. Meet., vol. 5511, Denver, CO: International Society for Optics and Photonics; 2004, p. 61-70. DOI: 10.1117/12.561724

[19] Cai Z, Smith NL, Zhang J-T, Asher SA. Two-dimensional photonic crystal chemical and biomolecular sensors. Analytical Chemistry. 2015;**87**:5013-5025. DOI: 10.1021/ac504679n

[20] Lee H-S, Lee E-H. Design of compact silicon optical modulator using photonic crystal MZI structure. 2008 5th IEEE Int Conf Gr IV Photonics 2008:308-310. DOI: 10.1109/GROUP4. 2008.4638182

[21] Viktorovitch P, Drouard E, Garrigues M, Leclercq JL, Letartre X, Rojo Romeo P, et al. Photonic crystals: Basic concepts and devices. Comptes Rendus Physique. 2007;**8**:253-266. DOI: 10.1016/j.crhy.2006.04.005

[22] Threm D, Nazirizadeh Y, Gerken M. Photonic crystal biosensors towards on-chip integration. Journal of Biophotonics. 2012;**5**:601-616. DOI: 10.1002/jbio.201200039

[23] YOLE. YOLE Reports abstract for EPIC. 2015

[24] MNT Gas Sensor Forum. MNT Gas Sensor Roadmap. 2006

[25] Sekhar PK, Brosha EL, Mukundan R, Garzon FH. Chemical sensors for environmental monitoring and homeland security. Electrochemical Society Interface. 2010:35-40. DOI: 10.1149/2.F04104if

[26] Ramboll Environ US Corporation. Technology Assessment Report: Air Monitoring Technology near Upstream Oil and Gas Operations. 2017

[27] Wetchakun K, Samerjai T, Tamaekong N, Liewhiran C, Siriwong C, Kruefu V, et al. Semiconducting metal oxides as sensors for environmentally hazardous gases. Sensors and Actuators B: Chemical. 2011;**160**:580-591. DOI: 10.1016/j.snb.2011.08.032

[28] OECD. Smart Sensor Networks : Technologies and Applications for Green Growth. 2009

[29] Tonouchi M. Cutting-edge terahertz technology. Nature Photonics. 2007;**1**:97-105. DOI: 10.1038/nphoton.2007.3

[30] Bigourd D, Cuisset A, Hindle F, Matton S, Fertein E, Bocquet R, et al. Detection and quantification of multiple molecular species in mainstream cigarette smoke by continuous-wave terahertz spectroscopy. Optics Letters. 2006;**31**:2356. DOI: 10.1364/OL.31.002356

[31] Lee DD. Environmental gas sensors. IEEE Sensors Journal. 2001;**1**:214-224. DOI: 10.1109/JSEN.2001.954834

[32] Han X, Naeher LP. A review of traffic-related air pollution exposure assessment studies in the developing world. Environment International. 2006;**32**:106-120. DOI: 10.1016/j.envint.2005.05.020

[33] Monn C. Exposure assessment of air pollutants: A review on spatial heterogeneity and indoor/outdoor/personal exposure to suspended particulate matter, nitrogen dioxide and ozone. Atmospheric Environment. 2001;**35**:1-32. DOI: 10.1016/S1352-2310(00)00330-7

[34] U.S. Environmental Protection Agency. American's Children and the Environment. 3rd ed. 2013

[35] Rothman LS, Gordon IE, Barbe A, Benner DC, Bernath PF, Birk M, et al. The HITRAN 2008 molecular spectroscopic database. Journal of Quantitative Spectroscopy and Radiation Transfer. 2009;**110**:533-572. DOI: 10.1016/j.jqsrt.2009.02.013

[36] Dinh T-V, Choi I-Y, Son Y-S, Kim J-C. A review on non-dispersive infrared gas sensors: Improvement of sensor detection limit and interference correction. Sensors and Actuators B: Chemical. 2016;**231**:529-538. DOI: 10.1016/j.snb.2016.03.040

[37] Mikuta R, Silinskas M, Bourouis R, Kloos S, Burte EP. Characterization of non-dispersive infrared gas detection system for multi gas applications. tm - Tech Mess. 2016;**83**:410-416. DOI: 10.1515/teme-2015-0010

[38] Painter O, Vučkovič J, Scherer A. Defect modes of a two-dimensional photonic crystal in an optically thin dielectric slab. Journal of Optical Society of America B. 1999;**16**:275. DOI: 10.1364/JOSAB.16.000275

[39] Nagahara K, Morifuji M, Kondow M. Optical coupling between a cavity mode and a waveguide in a two-dimensional photonic crystal. Photonics Nanostructures - Fundamental Applications. 2011;**9**:261-268. DOI: 10.1016/j.photonics.2011.04.011

[40] Hu J. Ultra-sensitive chemical vapor detection using micro-cavity photothermal spectroscopy. Optics Express. 2010;**18**:22174. DOI: 10.1364/OE.18.022174

[41] Vasiliev A, Malik A, Muneeb M, Kuyken B, Baets R, Roelkens G. On-chip mid-infrared Photothermal spectroscopy using suspended silicon-on-insulator microring resonators. ACS Sensors. 2016;**1**:1301-1307. DOI: 10.1021/acssensors.6b00428

[42] Wehrspohn RB, Schweizer SL, Gesemann B, Pergande D, Geppert TM, Moretton S, et al. Macroporous silicon and its application in sensing. Comptes Rendus Chimie. 2013;**16**:51-58. DOI: 10.1016/j.crci.2012.05.011

[43] Xu H, Wu P, Zhu C, Elbaz A, Gu ZZ. Photonic crystal for gas sensing. Journal of Materials Chemistry C. 2013;**1**:6087. DOI: 10.1039/c3tc30722k

[44] Śmigaj W, Gralak B. Validity of the effective-medium approximation of photonic crystals. Physical Review B. 2008;**77**:235445. DOI: 10.1103/PhysRevB.77.235445

[45] Li HH. Refractive index of silicon and germanium and its wavelength and temperature derivatives. Journal of Physical and Chemical Reference Data. 1980;**9**:561. DOI: 10.1063/1.555624

[46] Garín M, Trifonov T, Hernández D, Rodriguez A, Alcubilla R. Thermal emission of macroporous silicon chirped photonic crystals. Optics Letters. 2010;**35**:3348-3350. DOI: 10.1364/OL.35.003348

[47] Declair S, Meier T, Zrenner A, Förstner J. Numerical analysis of coupled photonic crystal cavities. Photonics Nanostructures - Fundamental Applications. 2011;**9**:345-350. DOI: 10.1016/j.photonics.2011.04.012

[48] Kolasinski KW. Etching of silicon in fluoride solutions. Surface Science. 2009;**603**:1904-1911. DOI: 10.1016/j.susc.2008.08.031

[49] Trifonov T, Rodríguez A, Marsal LF, Pallarès J, Alcubilla R. Macroporous silicon: A versatile material for 3D structure fabrication. Sensors and Actuators A: Physical. 2008;**141**:662-669. DOI: 10.1016/j.sna.2007.09.001

[50] Rönnebeck S, Ottow S, Carstensen J, Föll H. Crystal orientation dependence of macropore formation in n-Si with backside-illumination in HF-electrolyte. Journal of Porous Materials. 2000;**7**:353-356. DOI: 10.1023/A:1009639105357

[51] Carstensen J, Prange R, Föll H. A model for current-voltage oscillations at the silicon electrode and comparison with experimental results. Journal of the Electrochemical Society. 1999;**146**:1134-1140. DOI: 10.1149/1.1391734

[52] Föll H, Christophersen M, Carstensen J, Hasse G. Formation and application of porous silicon. Material Science and Engineering R Reports. 2002;**39**:93-141. DOI: 10.1016/S0927-796X(02)00090-6

[53] Carstensen J, Christophersen M, Hasse G, Föll H. Parameter dependence of pore formation in silicon within a model of local current bursts. Physica Status Solidi. 2000;**182**:63-69. DOI: 10.1002/1521-396X(200011)182:1<63::AID-PSSA63>3.0.CO;2-E

[54] Hilbrich S, Theiß W, Arens-Fischer R, Glück O, Berger M. The influence of the doping level on the optical properties of porous silicon. Thin Solid Films. 1996;**276**:231-234. DOI: 10.1016/0040-6090(95)08060-0

[55] Vega Bru D, Cardador Maza D, Trifonov T, Garin Escriva M, Rodriguez MA. The effect of absorption losses on the optical behaviour of macroporous silicon photonic crystal selective filters. Journal of Light Technology. 2016;**34**:1281-1287. DOI: 10.1109/JLT.2015.2503354

[56] Matthias S, Hillebrand R, Müller F, Gösele U. Macroporous silicon: Homogeneity investigations and fabrication tolerances of a simple cubic three-dimensional photonic crystal. Journal of Applied Physics. 2006;**99**:113102. DOI: 10.1063/1.2200871

[57] Segura D, Vega D, Cardador D, Rodriguez A. Effect of fabrication tolerances in macroporous silicon photonic crystals. Sensors Actuators, A Physics. 2017;**264**:172-179. DOI: 10.1016/j.sna.2017.07.011

[58] Langtry TN, Asatryan AA, Botten LC, de Sterke CM, McPhedran RC, Robinson PA. Effects of disorder in two-dimensional photonic crystal waveguides. Physical Review E 2003;**68**:26611. DOI: 10.1103/PhysRevE.68.026611

[59] Schilling J, Scherer A. 3D photonic crystals based on macroporous silicon: Towards a large complete photonic bandgap. Photonics Nanostructures - Fundamental Applications. 2005;**3**:90-95. DOI: 10.1016/j.photonics.2005.09.015

[60] Cardador D, Segura D, Vega D, Rodriguez A. Coupling defects in macroporous silicon photonic crystals. 2017 Spanish Conf. Electron Devices, Barcelona, Spain: IEEE; 2017, p. 1-3. DOI: 10.1109/CDE.2017.7905236

[61] Granados MJ. Fabricación de dispositivos fotónicos integrados. Bachelor's degree Thesis. Universistat Politècnica de Catalunya; 2017

[62] Joannopoulos JD, Johnson SG, Winn JN, Meade RD. Photonic Crystals: Molding the Flow of Light. 2.ª ed. Princeton University Press; 2011

New Approach to Mach-Zehnder Interferometer (MZI) Cell Based on Silicon Waveguides for Nanophotonic Circuits

Trung-Thanh Le

Additional information is available at the end of the chapter

http://dx.doi.org/10.5772/intechopen.76181

Abstract

In this chapter, we present a new scheme for Mach-Zehnder Interferometer (MZI) structure based on only one 4×4 multimode interference (MMI) coupler. We design the new MZI cell on the silicon on insulator (SOI) platform. The MZI based on directional coupler and 2×2 MMI coupler is also investigated in detail. The new MZI cell is a basic building block for photonic applications such as optical quantum gate, optical computing and reconfigurable processors. The numerical simulations show that our approach has advantages of compact size, ease of fabrication with the current complementary metal oxide semiconductor (CMOS) circuitry.

Keywords: multimode interference, silicon on insulator, multimode waveguide, directional coupler, finite difference time difference, finite difference method, modified effective index method

1. Introduction

Mach-Zehnder Interferometer (MZI) structure is a versatile component for photonic integrated circuits. A variety of photonic functional devices can be realized by using the MZI such as optical filter [1], add-drop multiplexer [2, 3], switch [4], modulator [5], sensor [6–8], tunable coupler [9, 10], signal transforms [11–13] and optical computing [14, 15], and so on.

In recent years, we have presented some approaches to realize optical signal processing based on MZI incorporated with microring resonators and multimode interference (MMI) coupler on a silicon on insulator (SOI) platform [16–20]. Silicon photonics is considered a key technology for next generation optical interconnects, optical computing, data center and communication systems due to its low power operation, compactness, scalability and compatibility with the CMOS process. The SOI platform has a large index contrast between the silicon core and the silicon oxide/air cladding, thus allowing for small bend waveguides and denser integration of photonic components. Silicon photonic devices are also being considered for wavelength-division multiplexing (WDM) metro and long haul network segments.

It was shown that MZI is a basic element (basic cell) for optical photonic circuits and quantum technologies [21]. In [21], by cascading 15 MZIs with 30 thermo-optic phase shifters, a single programmable optical chip has been implemented. In [22], the author has presented a self-aligned universal beam coupler based on MZI elements that can take an arbitrary monochromatic input beam and, automatically and without any calculations, couple it into a single-mode guide or beam. This device can be used for special optical applications such as automatic compensation for misalignment and defocusing of an input beam, coupling of complex modes or multiple beams from fibers or free space to single-mode guides, and retaining coupling to a moving source. By using recursive algorithm, any discrete finite dimensional unitary operator based on MZI elements can also be constructed [23]. In addition, it was shown that optical neural networks based on architecture of 56 MZIs with 213 thermo-optic phase shifter elements have been implemented successfully [24]. This photonic circuit has been fabricated in an SOI platform with OpSIS (University of Washington's Optoelectronic Systems Integration in Silicon) foundry. An optical neuromorphic computing on a vowel recognition dataset has been demonstrated experimentally.

Very recently, universal multiport photonic interferometers by means of arrangements of reconfigurable MZIs on SOI platform can implement any arbitrary unitary transformation between input and output optical modes [25]. These arbitrary transforms are essential to support advanced optical functions such as linear quantum optical gates and circuits, microwave photonics signal processors, spatial mode converters, data center connections and optical networking functionalities. The triangle, square arrangements of MZI elements for hardware architecture are similar to the design principles of the field programmable gate arrays (FPGAs) in electronics. The core concept is to use a large network of identical two-dimensional unit or lattice cells implemented by means of MZIs. With a proper MZI element lattice arrangement, the architecture can implement a variety of functional configurations by mapping the desired matrix to a selection of signal paths through the architecture. By introducing the phase shifter elements in two arms of the MZI, a reconfigurable architecture can be implemented.

In the literature, the MZI consists of two 3 dB 2 × 2 directional couplers or 2 × 2 MMI couplers linked though two waveguide arms. It was shown that directional coupler is very sensitive to the fabrication error [16]. The power coupling ratios can be controlled by adjusting the coupling length and/or the gap between the two waveguides of the directional coupler [16]. In practice, accurate fabrication of the gap requires very tight control of the fabrication process. Moreover, additional loss due to mode conversion loss has been found to be a problem [26]. In contrast, MMI-based devices often have large fabrication tolerance, wide operation bandwidth and compact size. As a result, it is attractive to realize new functional devices based on MMIs for photonic applications.

In this chapter, we present a new MZI element architecture based on only one 4 × 4 MMI coupler on an SOI platform. The phase shifter based on a PN junction, which use the plasma dispersion effect in silicon waveguides, is used. Our approach has advantages of compact size, ease of fabrication with the current CMOS circuit.

2. MZI cell based on directional couplers

2.1. Directional coupler type I

Figure 1 shows a general configuration of an MZI element based on two-directional couplers type I. Two arms of the MZI use two phase shifters with phase shifts ϕ_1 and ϕ_2. In our study, we use the PN junction operated in reverse bias, the depletion region is widened which lowers the overlap of the optical mode with charge carriers [5, 27]. As a result, the optical loss is low

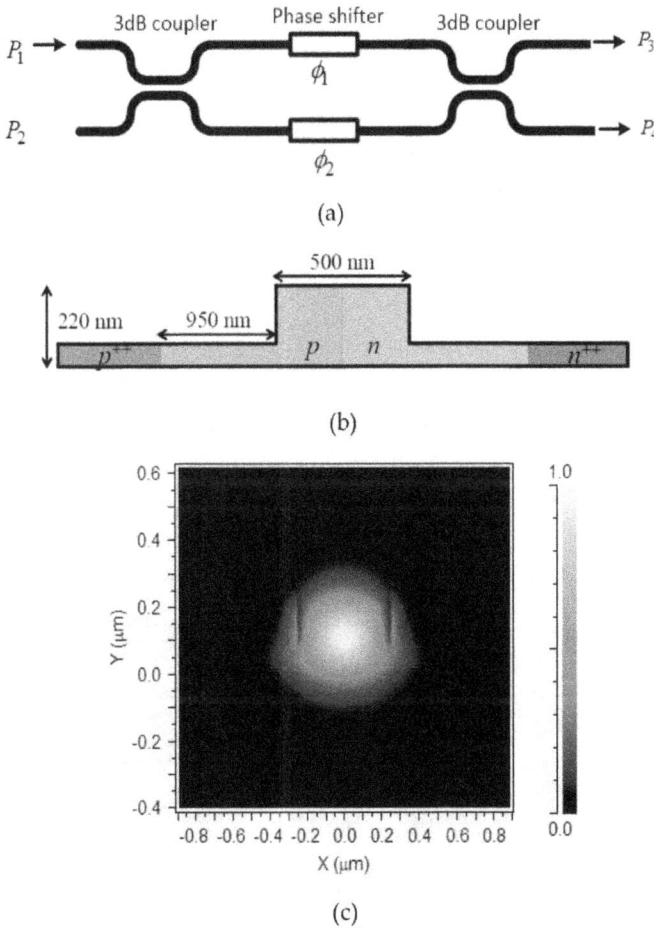

(a)

(b)

(c)

Figure 1. MZI cell based on directional coupler type I.

and the real part of the phase can be significantly increased compared with the forward bias. The change in index of refraction is phenomenologically described by Soref and Bennett model [28]. Here we focus on the central operating wavelength of around 1550 nm.

The change in refractive index Δn is described by:

$$\Delta n \, (\text{at 1550 nm}) \;=\; -8.8 \times 10^{-22} \Delta N - 8.5 \times 10^{-18} \Delta P^{0.8} \tag{1}$$

The change in absorption $\Delta \alpha$ is described by:

$$\Delta \alpha \, (\text{at 1550 nm}) \;=\; 8.5 \times 10^{-18} \Delta N + 6 \times 10^{-18} \Delta P \left[\text{cm}^{-1}\right] \tag{2}$$

where ΔN and ΔP are the free carriers concentrations of electrons and holes, respectively.

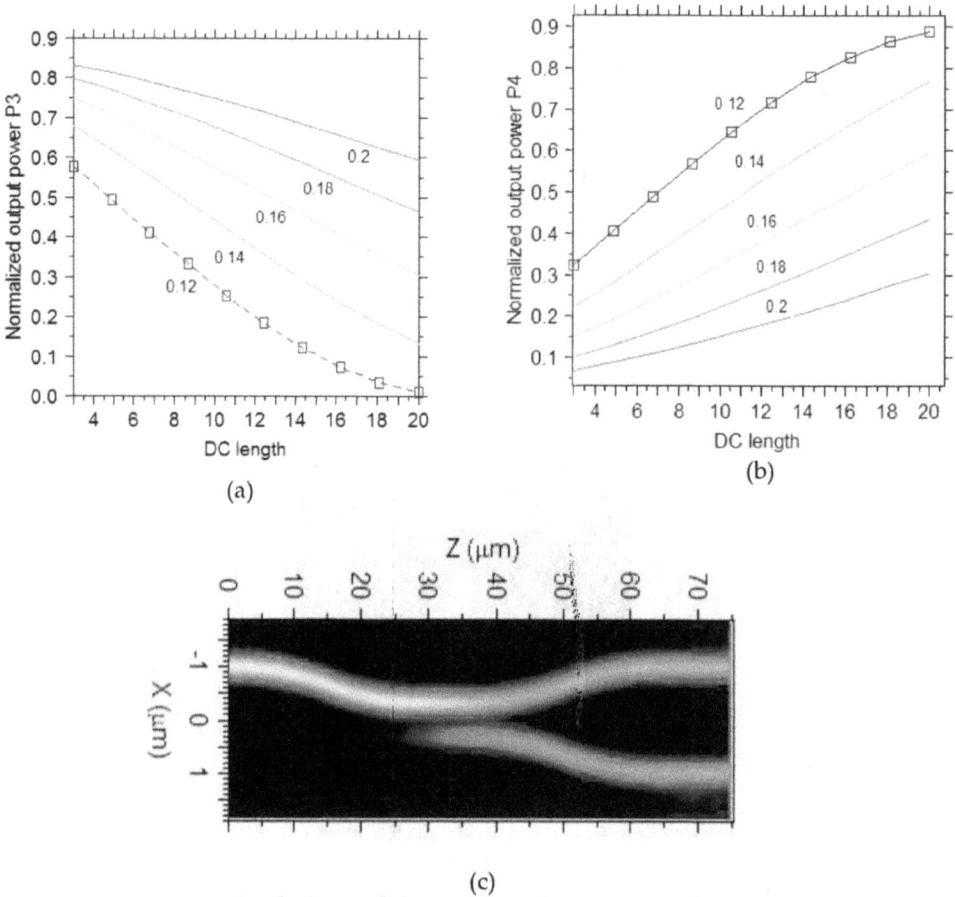

(a)

(b)

(c)

Figure 2. (a), (b) directional coupler design, type I for different separation between two waveguides g = 120, 140, 160, 180 and 200 nm and (c) field propagation at optimal length.

The directional coupler can be characterized by a matrix [29]:

$$M_{DC} = \begin{pmatrix} t & jr \\ jr & t \end{pmatrix} \tag{3}$$

where t and r are the transmission and coupling coefficients of the coupler. The transfer matrix method is used to analyze the working principle of an MZI element. The relationship matrix between the input and output fields of the MZI can be expressed as [30].

$$\mathbf{S} = \begin{bmatrix} t & jr \\ jr & t \end{bmatrix} \begin{bmatrix} e^{j\phi_1} & 0 \\ 0 & e^{j\phi_2} \end{bmatrix} \begin{bmatrix} t & jr \\ jr & t \end{bmatrix} = \begin{pmatrix} t^2 e^{j\phi_1} - r^2 e^{j\phi_2} & jrt(e^{j\phi_1} + e^{j\phi_2}) \\ jrt(e^{j\phi_1} + e^{j\phi_2}) & t^2 e^{j\phi_2} - r^2 e^{j\phi_1} \end{pmatrix} \tag{4}$$

where t and r are optical field transmission and cross-coupling coefficients, respectively. For a 3 dB coupler, $r = t = 0.707$ [6, 13], the transfer matrix of the MZI is expressed by:

$$\mathbf{S} = je^{j\frac{\Delta\phi}{2}} \begin{bmatrix} \sin\dfrac{\Delta\phi}{2} & \cos\dfrac{\Delta\phi}{2} \\ \cos\dfrac{\Delta\phi}{2} & -\sin\dfrac{\Delta\phi}{2} \end{bmatrix} \tag{5}$$

where $\Delta\phi = \phi_1 - \phi_2$ is the phase shift difference between two arms.

A type I directional coupler is shown in **Figure 1**. This coupler consists of two adjacent waveguides with sine shapes separated by a coupler gap g. The normalized output powers or t^2, r^2 at different directional coupler (DC) length for coupler gap g = 120, 140, 160, 180 and 200 nm are shown in **Figure 2**. From these simulation results, we can achieve the optimal length of the directional coupler for 50:50 coupling ratio or 3 dB coupler. **Figure 2(c)** shows a field propagation for a 3 dB coupler with g = 120 nm and optimal length of 5 µm.

2.2. Directional coupler type II

An MZI cell based on a type II directional coupler is shown in **Figure 3**. The coupler consists of two adjacent waveguides with sine shape and rectangular shape separated by a coupler gap g. The normalized output powers or t^2, r^2 at different directional coupler (DC) length for coupler gap g = 120, 140, 160, 180 and 200 nm are shown in **Figure 4**. From these simulation results, we can achieve the optimal length of the directional coupler for 50:50 coupling ratio or 3 dB coupler. **Figure 4(c)** shows a field propagation for a 3 dB coupler with g = 120 nm and optimal length of 2µm.

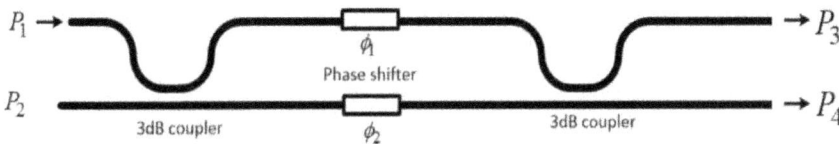

Figure 3. MZI cell based on directional coupler type II.

(a) Output power P3

(b) Output power P4

(c) Field propagation gap=120nm at optimal DC length of 2μm

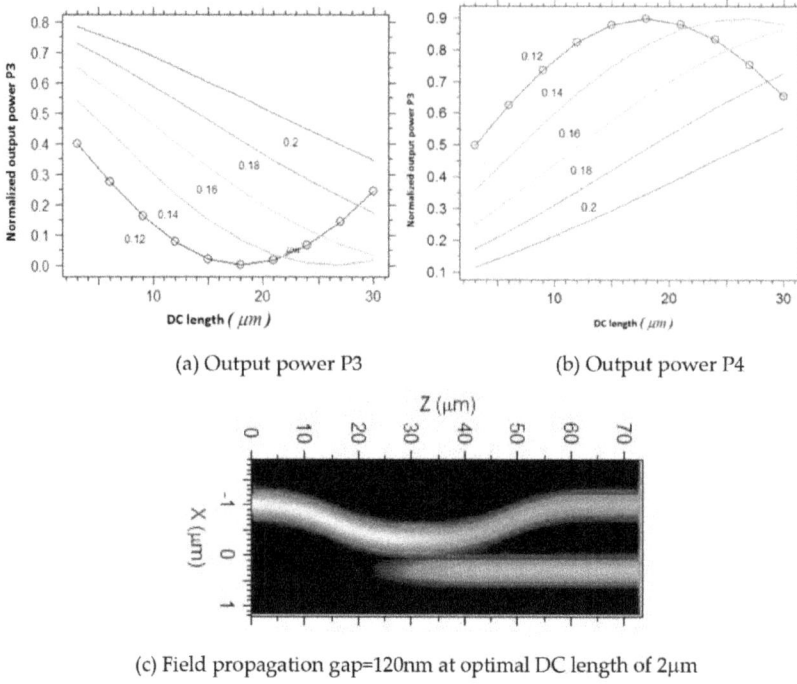

Figure 4. (a), (b) directional coupler design, type II for different separation between two waveguides g = 120, 140, 160, 180 and 200 nm and (c) field propagation at optimal length.

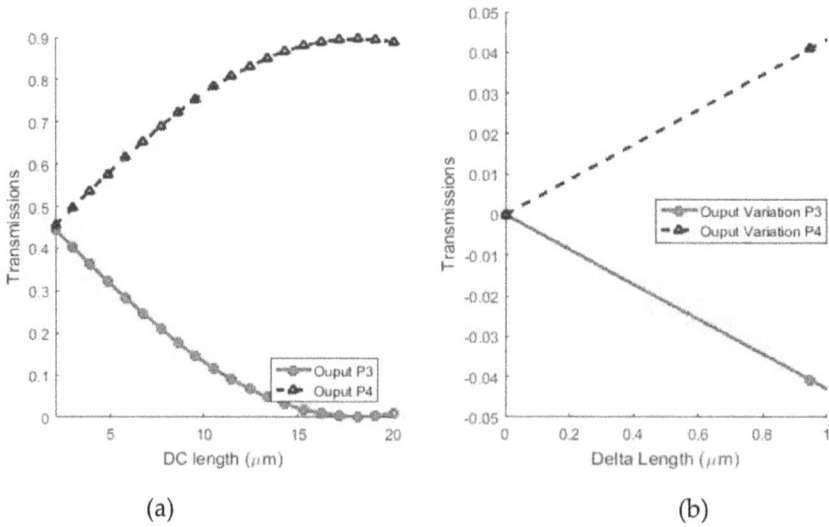

(a)

(b)

Figure 5. Transmissions of the directional coupler type II at different DC length.

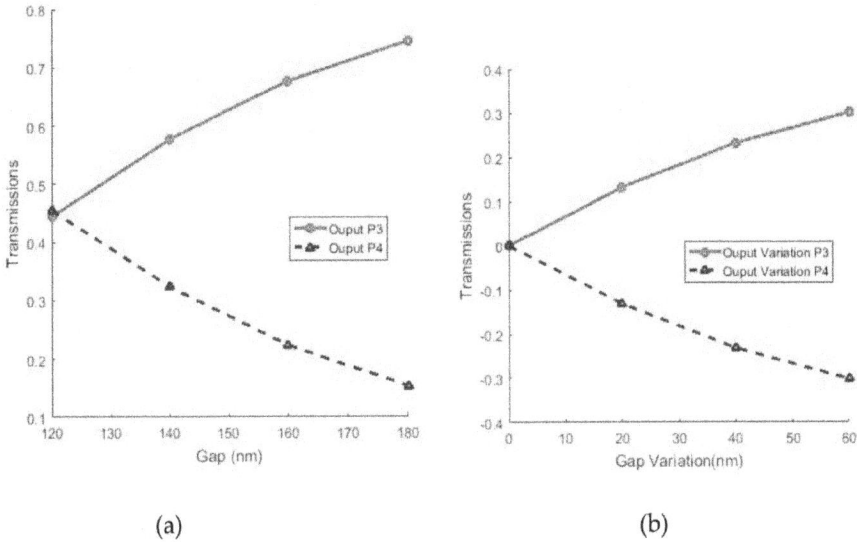

(a)   (b)

Figure 6. Transmissions of the directional coupler type II at different gap separation.

2.3. Fabrication tolerance analysis

Without loss of generality, we consider the directional coupler type II at the gap separation g = 120 nm for fabrication tolerance analysis. The fabrication tolerance of the directional coupler type I is similar to type II. The normalized output powers at the outputs are shown in **Figure 5**. We can see that a variation in length of ±100 nm results in a variation of 0.01 in normalized output powers. A change in a gap separation of ±10 nm will result in a change of 0.2 in normalized output power. As a result, the directional coupler type II is particularly sensitive to the fabrication error (**Figure 6**).

3. MZI cell based on 2 × 2 multimode interference coupler

The operation of an MMI coupler is based on the self-imaging principle or Talbot effect [31, 32]. The self-imaging shows that an input field is reproduced in single or multiple images at periodic intervals along the propagation direction of the MMI waveguide. The multimode waveguide is large enough to support a large number of modes.

There are two ways to create a 3 dB coupler based on MMI principle [20]: the first is the general interference (GI) mechanism which is independent of the modal excitation (we call MMI coupler type I in this chapter). The second is the restricted interference (RI) mechanism (MMI type II), in which excitation inputs are placed at some special positions so that certain modes are not excited.

3.1. MMI coupler type I

The MZI cell based on MMI couplers type I (GI-MMI) is shown in **Figure 7**. The MZI consists of two 3 dB MMI coupler. We have shown for the first time that a 3 dB GI-MMI coupler at the length of $1.5L_\pi$, where L_π is the beat length of the MMI coupler, can be expressed by a matrix [16]:

$$\mathbf{M}_{MMI,\,TypeI} = \frac{e^{j\phi_{GI}}}{\sqrt{2}} \begin{pmatrix} 1 & -j \\ -j & 1 \end{pmatrix} \tag{6}$$

where ϕ_{GI} is a constant phase of the coupler.

$$\mathbf{S} = \frac{e^{j\phi_{GI}}}{\sqrt{2}} \begin{pmatrix} 1 & -j \\ -j & 1 \end{pmatrix} \begin{pmatrix} e^{j\phi_1} & 0 \\ 0 & e^{j\phi_2} \end{pmatrix} \frac{e^{j\phi_{GI}}}{\sqrt{2}} \begin{pmatrix} 1 & -j \\ -j & 1 \end{pmatrix} = je^{j2\phi_{GI}} e^{j\frac{\Delta\varphi}{2}} \begin{pmatrix} \sin\dfrac{\Delta\varphi}{2} & \cos\dfrac{\Delta\varphi}{2} \\ \cos\dfrac{\Delta\varphi}{2} & -\sin\dfrac{\Delta\varphi}{2} \end{pmatrix} \tag{7}$$

The first step is to optimize the MMI sections: we use tapered waveguides with a length of 3 μm for access waveguides in order to improve the device performance. The multimode sections need to be wide enough to achieve good performance and to be spaced apart sufficiently to limit crosstalk between the adjacent access waveguides. By using the numerical simulations, we choose the width of the MMI waveguide is 3 μm. The three-dimensional beam propagation method (3D-BPM) is used to carry out the simulations for the device having a length of 52.2 μm. The aim of this step is to find roughly the positions which result in a power splitting of 50/50, that is, a 3 dB coupler. Then, the 3D-BPM is used to perform the simulations around these positions to locate the best lengths. The normalized output powers of the 2 × 2 GI-MMI coupler at different lengths of the couplers are plotted in **Figure 8**. The field propagations at the optimal length through the 3 dB GI-MMI coupler and the MZI cell are shown in

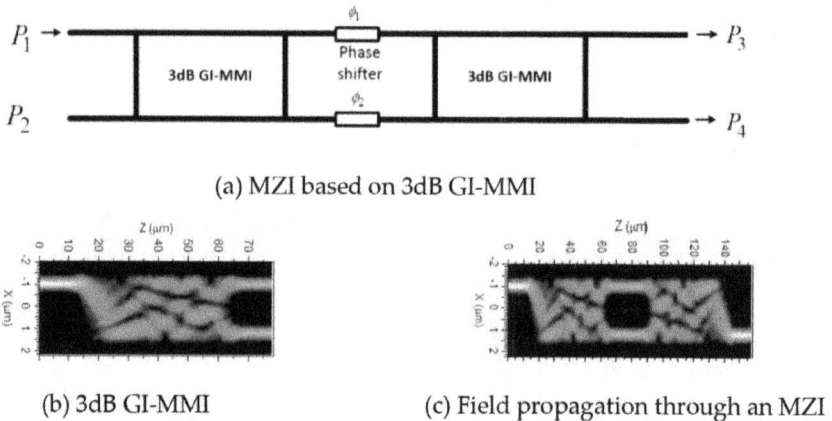

(a) MZI based on 3dB GI-MMI

(b) 3dB GI-MMI

(c) Field propagation through an MZI

Figure 7. MZI cell based on 3 dB MMI type I.

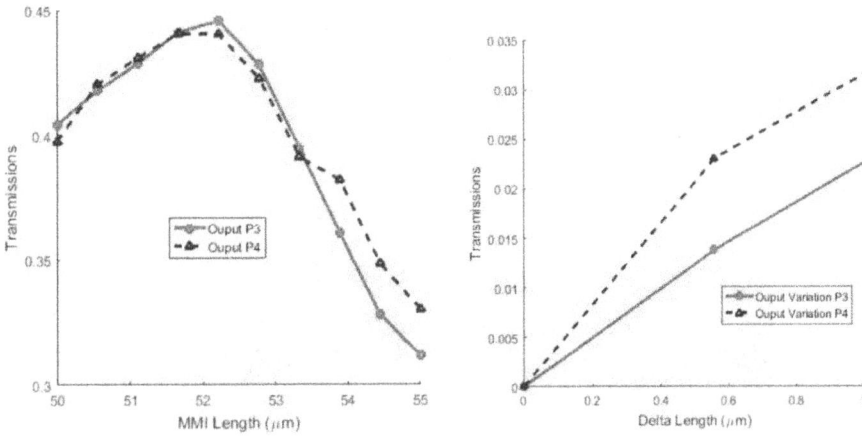

Figure 8. MZI cell based on 3 dB MMI type I.

Figure 7(b) and **(c)**. The simulations show that a variation in MMI length of ±100 nm will result in a change of 0.005 in normalized powers as shown in **Figure 8(b)**. Therefore, the MMI coupler has a much large fabrication tolerance compared with the directional coupler.

3.2. MMI coupler type II

The MZI cell based on MMI couplers type II (RI-MMI) is shown in **Figure 9**. We have shown that a 3 dB RI-MMI coupler at the length of $0.5L_\pi$, where L_π is the beat length of the MMI coupler, can be expressed by a matrix [16]:

$$\mathbf{M}_{MMI,\,TypeII} = \frac{e^{j\phi_{RI}}}{\sqrt{2}}\begin{pmatrix} 1 & j \\ j & 1 \end{pmatrix} \tag{8}$$

where ϕ_{RI} is a constant phase of the coupler.

$$\mathbf{S} = \frac{e^{j\phi_{RI}}}{\sqrt{2}}\begin{pmatrix} 1 & j \\ j & 1 \end{pmatrix}\begin{pmatrix} e^{j\phi_1} & 0 \\ 0 & e^{j\phi_2} \end{pmatrix}\frac{e^{j\phi_{RI}}}{\sqrt{2}}\begin{pmatrix} 1 & j \\ j & 1 \end{pmatrix} = je^{j2\phi_{RI}}e^{j\frac{\Delta\varphi}{2}}\begin{pmatrix} \sin\dfrac{\Delta\varphi}{2} & \cos\dfrac{\Delta\varphi}{2} \\ \cos\dfrac{\Delta\varphi}{2} & -\sin\dfrac{\Delta\varphi}{2} \end{pmatrix} \tag{9}$$

By using the numerical simulations, we choose the width of the MMI waveguide is 4 μm. The 3D-BPM is used to carry out the simulations for the device having a length of 33.4 μm. The normalized output powers of the 2 × 2 RI-MMI coupler at different lengths of the couplers are plotted in **Figure 10(a)**. The field propagations at the optimal length through the 3 dB RI-MMI coupler and the MZI cell are shown in **Figure 9(b)** and **(c)**. The simulations show that a variation in MMI length of ±100 nm will result in a change of 0.005 in normalized powers as shown in **Figure 10(b)**.

(a)

(b)   (c)

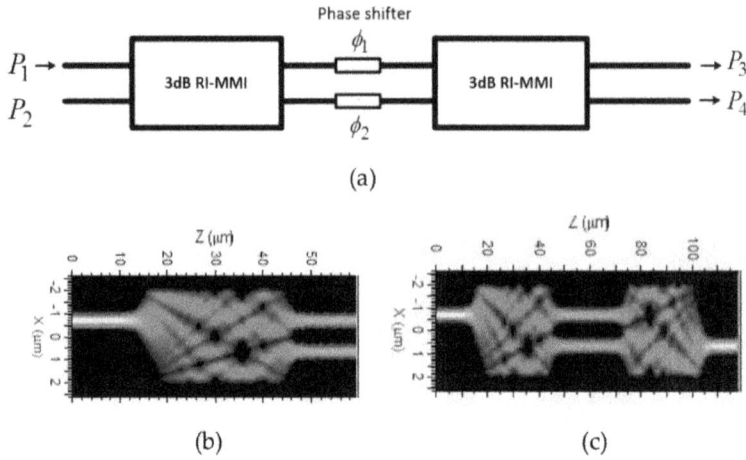

Figure 9. MZI cell based on 3 dB MMI type II.

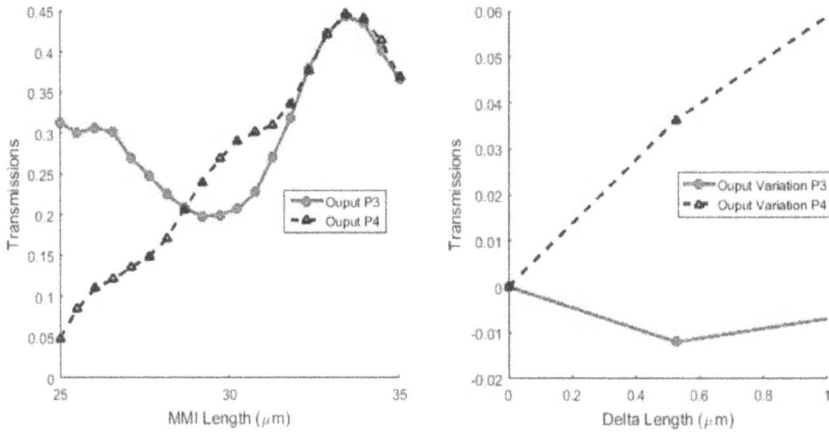

Figure 10. MZI cell based on 3 dB MMI type II.

4. MZI cell based on a 4 × 4 MMI coupler

Figure 11 shows the new scheme of our new proposed MZI cell based on only one 4 × 4 MMI coupler. We choose the width of the 4 × 4 MMI coupler is 6 μm. The optimal length of the MMI is calculated by the 3D-BPM [33]. We show that the optimal length is found to be 115.8 μm. **Figure 11(b)** and **(c)** shows the field propagation through the 4 × 4 MMI coupler at the optimal length for input signal at port 1 and port 2, respectively.

Figure 12 shows the normalized output powers at output port 1 and 4 while the input signal is at input port 1.

(a)

(b)

(c)

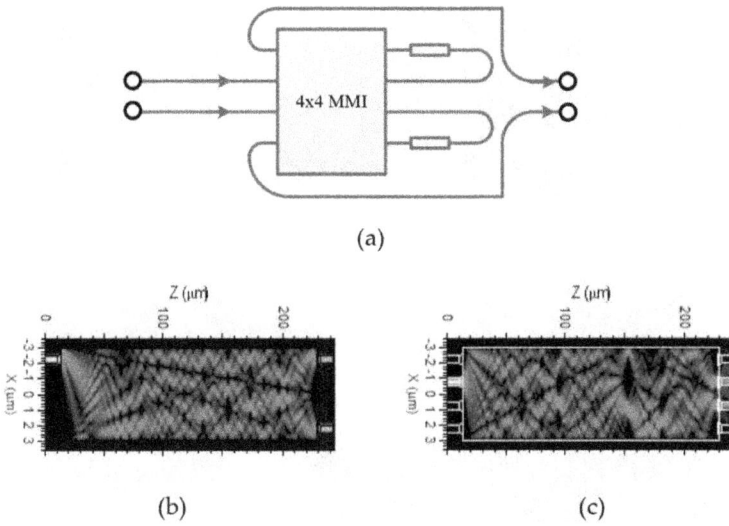

Figure 11. MZI cell based on 4 × 4 MMI coupler.

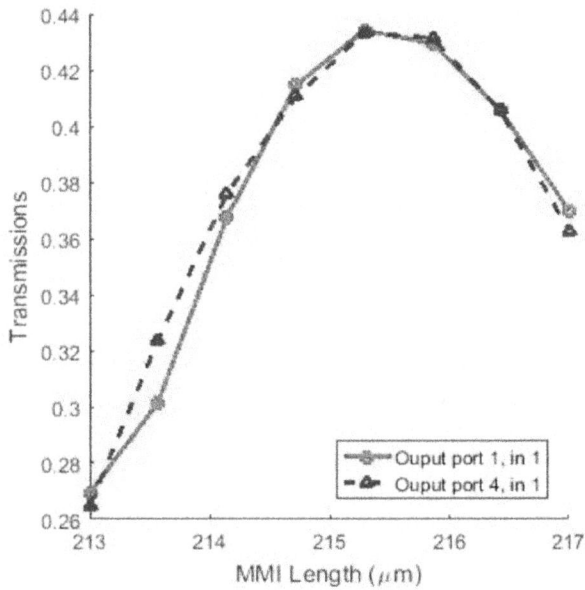

Figure 12. Normalized output powers at output port 1 and 4 of a 4 × 4 MMI coupler.

By some calculations, the MZI cell based on 4 × 4 MMI coupler can be expressed by a characterized matrix

$$\mathbf{S} = \frac{1}{\sqrt{2}}\begin{bmatrix} 1 & j \\ j & 1 \end{bmatrix}\begin{bmatrix} e^{j\Delta\varphi} & 0 \\ 0 & 1 \end{bmatrix}\frac{1}{\sqrt{2}}\begin{bmatrix} 1 & j \\ j & 1 \end{bmatrix} = e^{j\frac{\Delta\varphi}{2}}\begin{bmatrix} \tau & \kappa \\ \kappa^* & -\tau^* \end{bmatrix} \tag{10}$$

(a)

(b)

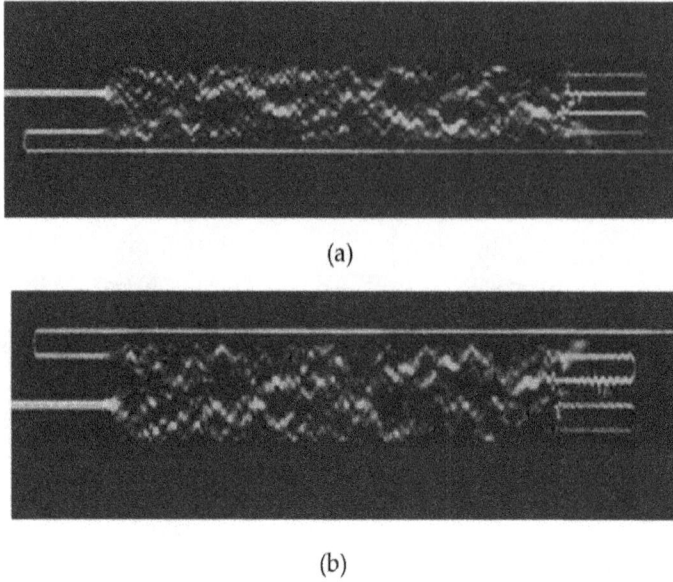

Figure 13. Optical field propagation through the coupler for input signal presented at port 2 and 3.

where $\tau = \sin\left(\frac{\Delta\varphi}{2}\right)$, and $\kappa = \cos\left(\frac{\Delta\varphi}{2}\right)$.

Finally, we use finite difference time difference (FDTD) method to simulate the proposed MZI cell and then make a comparison with the analytical theory. The proposed MZI cell has two feedback waveguides. Although the BPM has an advantage of fast simulations and it is widely used technique to simulate light propagation in slowly varying non-uniform guiding structures, it is not suitable for simulating the proposed MZI structure. The FDTD has a disadvantage of a time-consuming simulation, and it has very accurate results. In our FDTD simulations, we take into account the wavelength dispersion of the silicon waveguide. A light pulse of 15 fs pulse width is launched from the input to investigate the transmission characteristics of the device. The grid sizes $\Delta x = \Delta y = \Delta z = 20$ nm are chosen in our simulations for accurate simulations [34]. The FDTD simulations for the MZI cell are shown in **Figure 13**. The simulations show that the device operation has a good agreement with our prediction by analytical theory.

5. Waveguide mesh design with new MZI cell

The new proposed MZI cell is a basic building block for mesh design to produce new functional devices for photonic applications. For example, by using the hexagonal mesh shown in

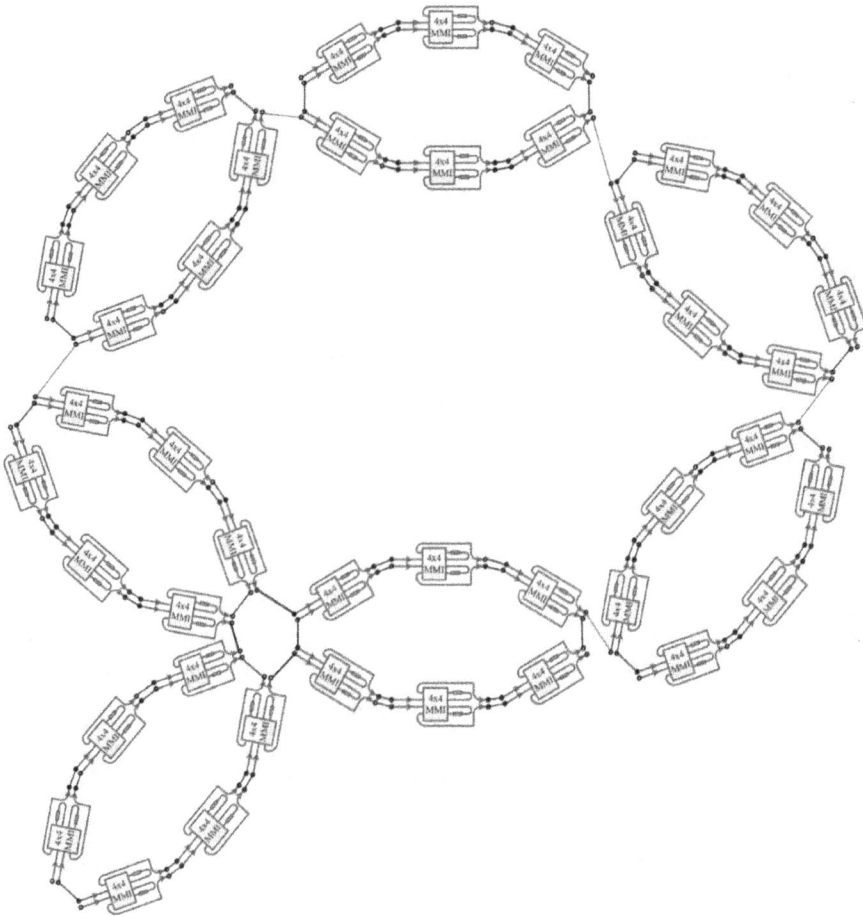

Figure 14. Reconfigurable mesh design based on new MZI cell.

Figure 14, optical filters and switches can be achieved by properly controlling the phase shifters.

6. Conclusions

We presented a new compact MZI cell based on silicon on insulator waveguides. The structure requires only one 4×4 multimode interference coupler. The PN junction waveguide is used to achieve reconfigurable devices. The device operation has been verified by using the FDTD. This MZI cell can be useful for complex mesh designs applied to optical applications.

Author details

Trung-Thanh Le

Address all correspondence to: thanh.le@vnu.edu.vn

International School (VNU-IS), Vietnam National University (VNU), Hanoi, Vietnam

References

[1] Agrawal GP. Fiber Optic Communication Systems. New York: John Wiley & Sons; 2002

[2] Le T-T, Cahill L. Microresonators based on 3×3 restricted interference MMI couplers on an SOI platform. In: IEEE LEOS Annual Meeting Conference Proceedings (LEOS 2009). Belek-Antalya, Turkey; October 4–8, 2009. pp. 479-480

[3] Le T-T, Cahill L. The design of 4×4 multimode interference coupler based microring resonators on an SOI platform. Journal of Telecommunications and Information Technology. 2009;**2**:98-102

[4] Le TT, Cahill LW. The design of wavelength selective switches and filters based on SOI microring resonators. In: 2007 Asia-Pacific Conference on Communications; 2007. pp. 3-5

[5] Le D-T, Nguyen N-M, Le T-T. Development of PAM-4 signaling for high performance computing, supercomputers and data center systems. Journal of Science and Technology on Information and Communications. 2017;**1**:34-38

[6] Le T-T. Realization of a multichannel chemical and biological sensor using 6×6 multimode interference structures. International Journal of Information and Electronics Engineering. 2011;**2**:240-244

[7] Le T-T. Microring resonator based on 3×3 general multimode interference structures using silicon waveguides for highly sensitive sensing and optical communication applications. International Journal of Applied Science and Engineering. 2013;**11**:31-39

[8] Le T-T. Two-channel highly sensitive sensors based on 4×4 multimode interference couplers. Photonic Sensors. 2017;**7**:357-364, 2017/12/01

[9] Le TT, Cahill LW, Elton DM. Design of 2×2 SOI MMI couplers with arbitrary power coupling ratios. Electronics Letters. 2009;**45**:1118-1119

[10] Le T-T. Arbitrary power splitting couplers based on 3×3 multimode interference structures for all-optical computing. International Journal of Engineering and Technology. 2011;**2**:565-569

[11] Le T-T. Realization of all-optical type i discrete cosine and sine transforms using multimode interference structures. International Journal of Microwave and Optical Technology (IJMOT). 2012;**7**:127-134

[12] Le T-T. All-optical Karhunen–Loeve transform using multimode interference structures on silicon nanowires. Journal of Optical Communications. 2011;**32**:217-220

[13] Le T-T. The design of optical signal transforms based on planar waveguides on a silicon on insulator platform. International Journal of Engineering and Technology. 2010;**2**:245-251

[14] Le T-T. Arbitrary power splitting couplers based on 3×3 multimode interference structures for all-optical computing. International Journal of Engineering and Technology. 2011;**3**:565-569

[15] Le D-T, Le T-T. Coupled resonator induced transparency (CRIT) based on interference effect in 4×4 MMI coupler. International Journal of Computer Systems (IJCS). 2017;**4**:95-98

[16] Le T-T. Multimode Interference Structures for Photonic Signal Processing. Germany: LAP Lambert Academic Publishing; 2010

[17] Cahill LW, Le TT. MMI devices for photonic signal processing. In: 9th International Conference on Transparent Optical Networks (ICTON 2007). Rome, Italy; July 1–5, 2007. pp. 202-205

[18] Cahill LW, Le TT. Photonic signal processing using MMI elements. In: 10th International Conference on Transparent Optical Networks (ICTON 2008); Athens, Greece; 22-26 June 2008. pp.114-117

[19] Le TT, Cahill LW. The modeling of MMI structures for signal processing applications. In: Greiner CM, Waechter CA, editors. Integrated Optics: Devices, Materials, and Technologies XII. Proceedings of the SPIE. Vol. 6896. San Jose, California, United States: Society of Photo-Optical Instrumentation Engineers (SPIE); pp. 68961G-68961G-7, 03/2008

[20] Le TT, Cahill L. All-optical signal processing circuits using silicon waveguides. In: The 7th International Conference on Broadband Communications and Biomedical Applications. Melbourne, Australia; ; November 21–24, 2011. pp. 167-172

[21] Carolan J, Harrold C, Sparrow C, et al. Universal linear optics. Science. 2015;**349**:711

[22] Miller DAB. Self-aligning universal beam coupler. Optics Express. 2013;**21**:6360-6370

[23] Reck M, Zeilinger A, Bernstein HJ, et al. Experimental realization of any discrete unitary operator. Physical Review Letters. 1994;**73**:58-61

[24] Shen Y, Harris NC, Skirlo S, et al. Deep learning with coherent nanophotonic circuits. Nature Photonics. 2017;**11**:441

[25] Perez D, Gasulla I, Fraile FJ, et al. Silicon photonics rectangular universal interferometer. Laser & Photonics Reviews. 2017;**11**:1700219

[26] Xia F, Sekaric L, Vlasov YA. Mode conversion losses in silicon-on-insulator photonic wire based racetrack resonators. Optics Express. 2006;**14**:3872-3886

[27] Baehr-Jones T, Ding R, Liu Y, et al. Ultralow drive voltage silicon traveling-wave modulator. Optics Express. 2012;**20**:12014-12020

[28]   Emelett SJ, Soref R. Design and simulation of silicon microring optical routing switches. IEEE Journal of Lightwave Technology. 2005;**23**:1800-1808

[29]   Heebner J, Grover R, Ibrahim T. Optical Microresonators: Theory, Fabrication, and Applications. London: Springer; 2008

[30]   Le T-T, Cahill L. Generation of two Fano resonances using 4 × 4 multimode interference structures on silicon waveguides. Optics Communications. 2013;**301-302**:100-105

[31]   Bachmann M, Besse PA, Melchior H. General self-imaging properties in N × N multimode interference couplers including phase relations. Applied Optics. 1994;**33**:3905

[32]   Soldano LB, Pennings ECM. Optical multi-mode interference devices based on self-imaging: Principles and applications. IEEE Journal of Lightwave Technology. 1995;**13**:615-627

[33]   Le T-T. An improved effective index method for planar multimode waveguide design on an silicon-on-insulator (SOI) platform. Optica Applicata. 2013;**43**:271-277

[34]   Le D-T, Nguyen M-C, Le T-T. Fast and slow light enhancement using cascaded microring resonators with the Sagnac reflector. Optik – International Journal for Light and Electron Optics. 2017;**131**:292-301

Chapter 4

Experimental Study of Porous Silicon Films

Salah Rahmouni and Lilia Zighed

Additional information is available at the end of the chapter

http://dx.doi.org/10.5772/intechopen.74479

Abstract

In the present study, porous silicon films were prepared on N- and P-type silicon wafer (100) crystallographic orientations. We have investigated the influence of the different anodization parameters and silicon wafers on the properties of the obtained porous silicon layer such as thickness and porosity. The reflectance measurement of the prepared samples has presented reduction of reflection due to the porous layers and suggests the antireflective character of the realized porous layer.

Keywords: porous silicon, antireflective coating, electrochemical anodization, solar cell, HF

1. Introduction

The increasing need of energy induces a strong greenhouse gas emission since energy production is mainly achieved by fossil fuel combustion. This gas excess has a harmful effect on the life on earth. To satisfy the increasing demand in energy, without altering our environment, it is indispensable to recourse to clean and renewable energies. Solar energy is one of the most promising renewable energy sources. Photovoltaic electricity is obtained by direct transformation of sunlight to electricity by means of solar cell. During the last few decades, photovoltaic market has an increasable progress. This leads to a growing research activity based on new materials and devices for obtaining more efficient solar cells with low cost. The reduction of optical energy losses is one of the most important factors in manufacturing high-efficiency silicon solar cells [1]. After its discovery in 1956 by Uhlir [2], porous silicon has attracted more intentions due to their interesting properties like antireflective coating, improving photovoltaic conversion efficiency [3–7]. Furthermore, its important specific surface has attracted earlier a technological interest in photoluminescence at room temperature and electroluminescence [8].

Electrochemical etching is considered as the most appropriate method to produce homogenous porous silicon [9, 10]. The aim of this study is the comparison between two types of porous silicon (P and N) generated with the above electrochemical method in order to see the effect of several experimental parameters (current density, anodization time, and hydrofluoric acid "HF" concentration) on the porosity, thickness, and antireflective activity, respectively.

The present work deals with the production of porous silicon by electrochemical way, to use it as antireflective coating for solar cell. We have studied the influence of the experimental parameters such as anodization time, current, and substrate type N or P on the final porous layer thickness, porosity, and antireflective activity.

2. Experimental details

PS layer was fabricated on P-type and N-type single crystal silicon wafers with a (100) crystallographic orientation and a resistivity of 3–5 Ωcm. After a standard cleaning process, a good ohmic aluminum contact has been evaporated onto the back of samples. Anodization was then performed in a [2, 3] volume ratio of a solution composed of 40% HF and 98% ethanol.

The electrochemical anodization of silicon is achieved by a homemade system; the experimental setup is presented in **Figure 1**.

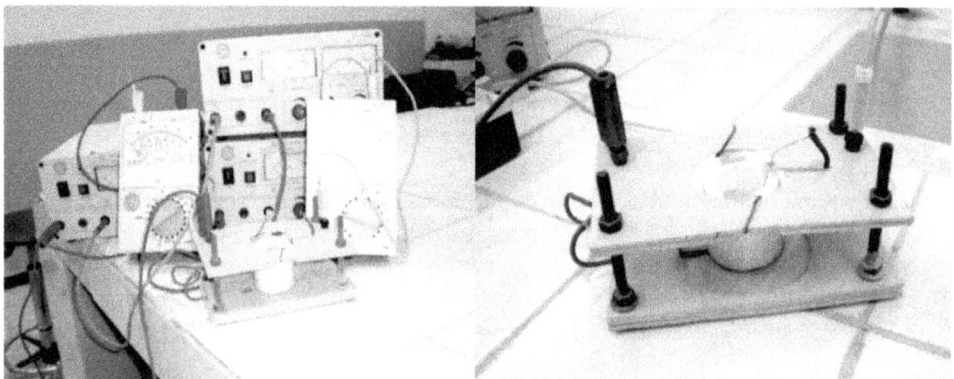

Figure 1. Photograph of used experimental device for Si substrate anodization.

3. Experimental protocol

3.1. Design of the anodizing platform

In order to produce porous silicon layers with a diameter of ≈10 mm, we proceeded the design of an anodizing nacelle with a circular opening.

Figure 2 shows the schematic diagram of the anodizing device consisting of the anodizing nacelle, along with a stabilized power supply that provides a constant current. This anodizing cell uses a metal electrical contact on the back side of the silicon wafer, isolated from the HF/ethanol solution by an HF inert O-ring. Thus, only the front face is exposed to the electrolyte attack. Obviously, the diameter of the O-ring controls the diameter of the obtained porous silicon stain, which remains valid when the edge effects are neglected.

3.2. Preparation of the substrate

The porous silicon samples were made by using the electrochemical anodizing method. This method is widely used, providing homogeneous porous layers, the porosity and the thickness of the elaborate layer are so controllable.

The silicon (Si) substrates used are monocrystalline, oriented (100), and (111) P and N type. Before anodization process, and in order to ensure a good electrical contact, an aluminum plate is placed on the back face of the substrate.

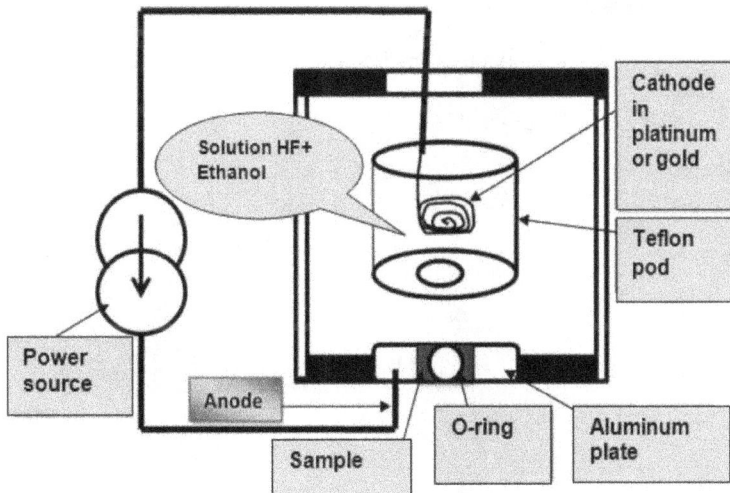

Figure 2. Diagram of the porous silicon manufacturing system.

3.3. Cleaning protocol

Before preparation, the samples must be well cleaned accordingly to the following protocol presented (**Figure 3**) in the synoptic diagram of the cleaning protocol.

3.4. Anodizing

3.4.1. Mounting of elaboration

The porous silicon layers (PS) are obtained after electrochemical etching in solution of hydrofluoric acid (HF) and ethanol (C_2H_5OH). We note that the role of ethanol is to standardize the porous layer and to minimize the formation of hydrogen bubbles at the surface, which homogenizes the porous layer [11, 12].

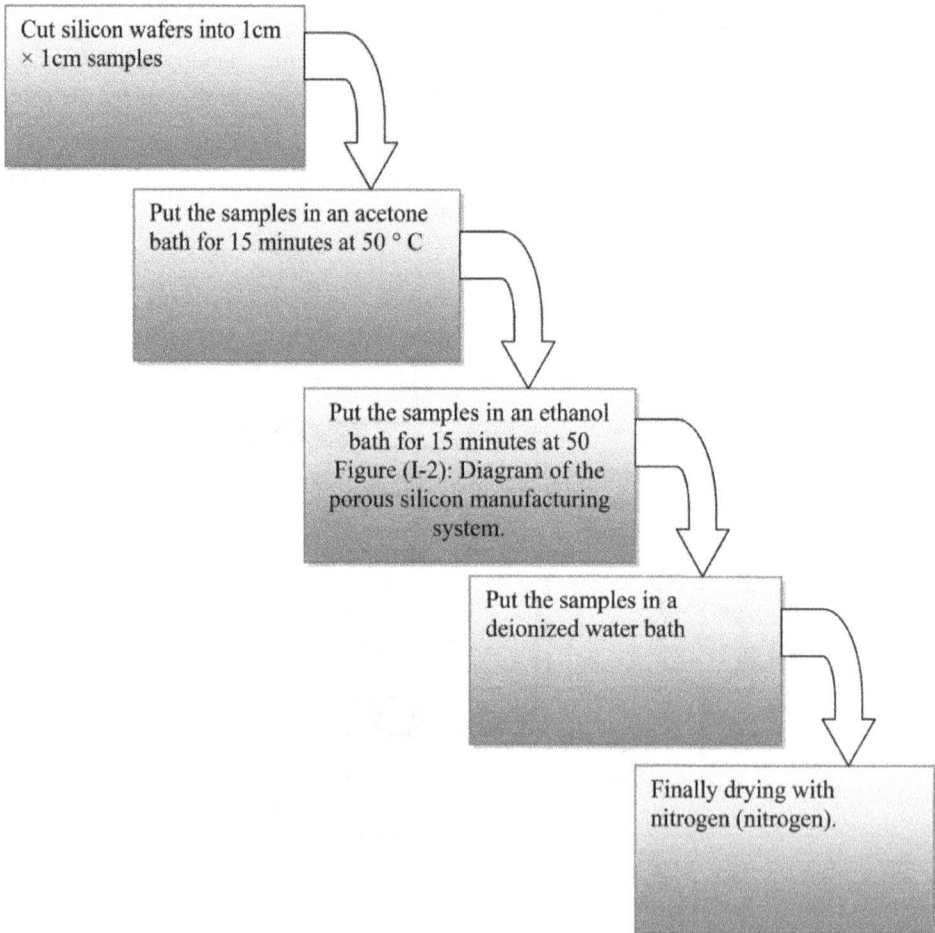

Figure 3. Synoptic diagram of the cleaning protocol.

The electrochemical dissolution is carried out in an impermeable material known as Teflon cell, by using the electrolyte of hydrofluoric acid solution having a high corrosive property. The cathode used in our electrochemical reactions was a circular gold or platinum grid. The Si substrate playing the role of anode is placed vertically against a circular orifice having a surface of 0.64 cm² in contact with the electrolyte. The shape of the cathode makes it possible to ensure a uniform distribution of the electric field lines, and subsequently the electrochemical etching of the substrate becomes homogeneous (see **Figure 2**). To ensure the reproducibility of the morphology of the layers and to control the porosity and the thickness, the constant-current dissolution which circulates between the two electrodes has been performed.

3.4.2. Anodizing parameters

Electrolyte concentration, current density, type and doping rate of silicon, electrolyte temperature, and illumination [12–16] are essential parameters for production of porous silicon.

i. Type and doping rate: A nano- and macroporous PS distribution was carried out on P- and N-type Si substrates. The spongy structure obtained is composed of nanocrystallites with sizes varied between 1 and 5 nm, separated by nanopores of the same type dimensions, for P-type substrates [11, 17–19], and macropores for N-type substrates [20, 21].

ii. Composition of the electrolyte: The electrolyte is composed of 40% HF and ethanol. Vial and Derrien [22] showed that porosity decreases with increasing concentration at a constant current density. In our case, we used several volume ratios for the different elaborations.

Silicon quality	Samples	Anodizing current density (mA/cm²)	Anodizing time (s)	Solution volume ratio (HF:C$_2$H$_5$OH)
Solar	A1(P100)	15	60	1:1
	A2(P100)	15	120	1:1
	A3(P100)	15	180	1:1
	A4(P100)	15	240	1:1
Solar	B1(P100)	5	180	1:1
	B2(P100)	10	180	1:1
	B3(P100)	15	180	1:1
	B4(P100)	20	180	1:1
Electronic	C1(P100)	30	120	3:2
	C2(P100)	10, 15, 18, 20, 30, 35, 54, 70	60, 120, 180, 240, 300, 360	2:3
	C3(N100)	10, 18, 30, 35, 54, 70, 214		2:3, 1:3
	C4(N111)	10, 20, 35, 106, 140	60, 120, 180, 240, 300, 360	2:3
			60, 120, 180, 240	
			300, 360	

Table 1. The conditions of anodization of the used substrates.

iii. Anodizing current: Note that the anodizing current and the concentration of the electrolyte have opposite effects in the formation of pores [23]. For a given concentration of the electrolyte, the porosity increases as a function of the current density. In our case, the anodizing conditions are detailed in **Table 1**.

Table 1 shows detailed data of the training conditions relating to the various samples of porous silicon produced.

4. Properties of substrates

In our work, we have used substrates of monocrystalline silicon types N and P, whose characteristics are presented in **Table 2** below.

Substrates	Dopant	Resistivity Ωcm	Orientation
N	Phosphore	(5–7)	100
N	Phosphore	(3–5)	111
P	Bore	(3–5)	100
P	Bore	(0.01–2)	100

Table 2. Characteristic of the used substrates.

5. The I (V) characteristic

The (I-V) characteristic of the electrolyte semiconductor junction depends on the nature of the semiconductor substrate as well as the ionic and molecular species present in the electrolyte. The application of an electrical potential to the silicon bathed in a solution of HF (see **Figure 1**) induces a measurable current circulating through the system.

At the silicon/electrolyte interface, the electronic charge carriers in the silicon of ionic form pass into the solution. This conversion is carried out by means of a redox reaction.

The value of the applied potential and the reaction taking place at the interface influences on the formation of porous silicon.

For the study of the I (V) characteristic the electrolytic solution which we used is formed of the following volume ratio fractions: 2 HF:3 ethanol.

In our experimental work, we used absolute ethanol and a hydrofluoric acid concentration of 40%. The previous proportions, i.e., 2 HF:3 ethanol, reduce the concentration [HF] in the resulting electrolyte to 16%.

Figure 4 shows the I (V) characteristic of the electrochemical anodization of a sample of monocrystalline silicon types (a) N (100), (b) N (111), and (c) P (100).

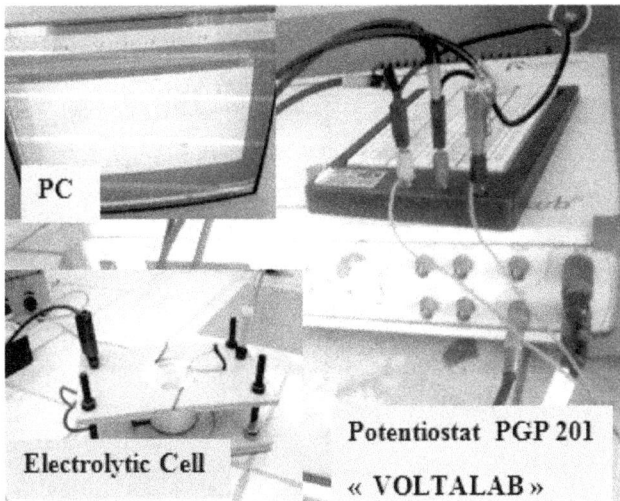

Figure 4. Mounting used for sample preparation.

There are three areas that characterize the anodizing process, the I (V) curves, showing the existence of an I_{ps} current delineating the formation and electropolishing zones.

5.1. Porosification zone (I < I_{ps})

In our work and using N-type substrates, the obtained I (V) characteristics show a possibility to prepare porous silicon with high currents. The porosification zone remains the same with somewhat high values of current. This is due to the high value of the resistivity on the one hand and also the nature of the substrates on the other hand (N type), in addition to the fact that the experiment takes place in the dark and at room temperature. Under these conditions, the intrinsic concentration of holes is too low to form pores, so it is necessary to generate holes by applying a high potential [24]. The layers are obtained using the following conditions.

In the case of the P-type substrate, the concentration of the holes on the surface is greater than that of the fluorine ions, so anodization leads to the formation of pores.

The Jps depends mostly on the composition of the HF solution in ethanol and little on the substrate.

5.2. Electropolishing zone (I > I_{ps})

According to the literature, this zone appears when the anode potentials are high [25]. For N-type substrates, when I > I_{SP} the limiting factor becomes the diffusion of the ionic species in the electrolyte, the holes then in excess at the bottom of the pores penetrate the porous structure which gradually generates the total dissolution of the latter [26].

In the case of P-type substrate, the density of holes on the silicon surface is important. Etching is limited by the diffusion of F-fluorine ions.

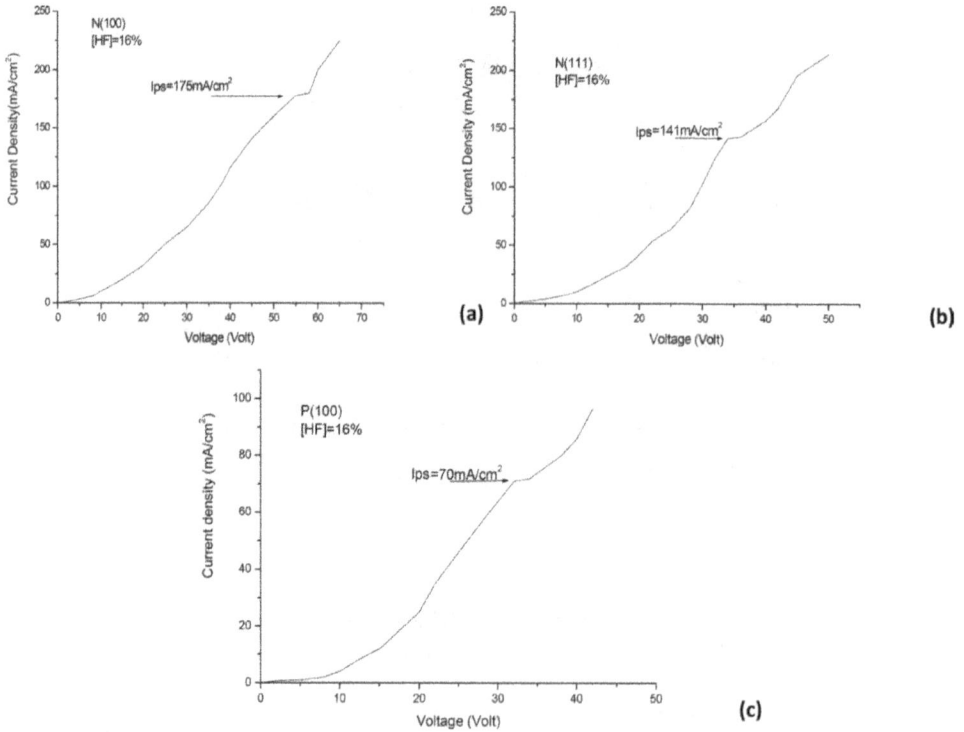

Figure 5. I (V) characteristic of silicon substrate electrochemical anodization types (a) N (100), (b) N (111), and (c) P (100).

The latter are attracted by the electric field located on the surface defects; the engraving in these places will be predominant tending to smooth the surface.

According to Ozanam and Chazalviel, electropolishing exists with an oxide layer that forms on the silicon surface for $I > I_{SP}$ currents [25].

5.3. Transition zone

If the anode potentials are at an intermediate level, there is a so-called transition zone. Since the morphology of the resulting surface is porous in nature, pore size increases rapidly with increasing potentials, leading to electropolishing of the surface.

This zone is characterized by a peak current corresponding to the formation of an oxide layer necessary for the surface electropolishing reaction [27].

6. Measurement of the thickness

6.1. Influence of anodizing time on thickness

In regard to the thickness measurement by profilometry, we studied the influence of the various anodizing parameters.

The samples studied are obtained using an electrolytic solution of a concentration

[HF] = 16%, in which the current density j is 35 mA/cm², for the silicon P (100). For silicon N (100), we used a current density of 54 mA/cm².

From **Figure 6** the thickness of the porous layer increases linearly with the anodizing time, for a current density and a given RF concentration. The number of dissolved silicon atoms is therefore directly proportional to the amount of charge exchanged (Q = j × the dissolution time) showing that the dissolution valence is invariant in time.

In the limit of the porous silicon formation regime, similar behaviors are observed, whatever the anodization current and the HF concentration [25–28].

6.2. Influence of the current density on the thickness

To know the influence of the current density on the thickness of the porous layer and on the etch rate, we set the attack time at 2 min for the three types of silicon that are available to us. For silicon types N (111) and P (100), we used an electrolytic solution concentrated at 16% HF. For the substrate type N (100), we have the same solution concentrated at 17.1%. The results obtained are presented by the graphs of **Figure 7**.

Figure 7 shows the variation of the thickness of the porous layer and the etch rate versus anodization current density for silicon substrates: (a) type N (111), (b) type P (100), and (c) type N (100). We note that the thicknesses of the obtained layers and the etch rates are increasing in a function of the current density [28].

For the N-type substrate (111), the thickness (the etch rate) of the porous layers increases linearly from 1.4 μm (0.7 μm/min) for j varying from 18 mA/cm² up to 106 mA/cm². For N- (100)

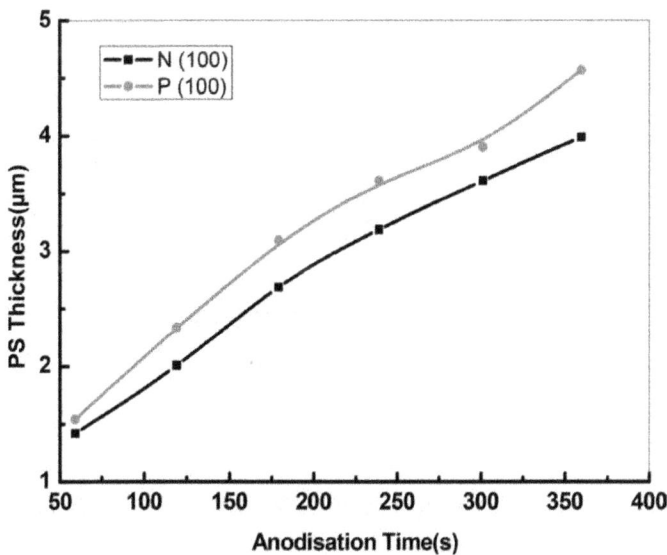

Figure 6. Variation of PS layer thickness as a function of anodization time in N-type Si (100) and P-type Si (100) substrates. The used conditions are [HF] = 16% and current density of 35 mA/cm².

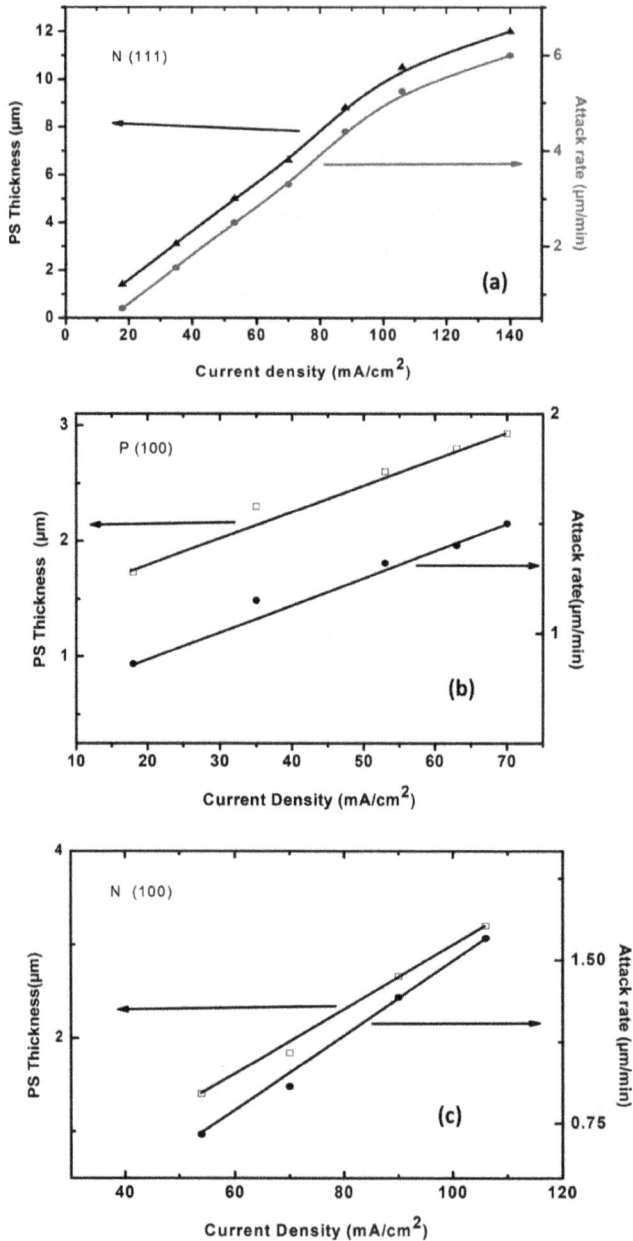

Figure 7. PS layers thickness variation and etch rate variation versus current density in (a) N-type Si (100) and (b) P-type Si (100) wafers.

and P- (100) type substrates, we note that the variation in thickness (etch rate) versus current density is linear.

For type N (100), the variation increases from 1.4 μm (0.7 μm/min) for j = 54 mA/cm² to 3.19 μm (1.68 μm/min) for j = 106 mA/cm². In the case of the substrate type P (100), the variation increases from 1.7 µm (0.12 µm/min) for j = 18 mA/cm² up to 2.9 µm (1 µm/min) for j = 70 mA/cm².

7. Determination of porosity

One of the important features of porous silicon is the degree of porosity, i.e., the percentage of void in the porous silicon volume [29]. In our work, the determination of the porosity of the studied samples was carried out by the gravimetric method.

In our study, the porous silicon layers are prepared at room temperature and without illumination. The electrolytic solution used is prepared from 40% concentrated hydrofluoric acid diluted in absolute ethanol. To know the variation of the porosity as a function of the current density, we fixed the anodization time at 2 min.

7.1. Influence of current density on porosity

Our study is made on silicon P (100) and N (100).

The variations in porosity as a function of the current density of the samples prepared are shown in **Figure 8**.

Figure 8 shows that porosity increases as a function of current density [28]. It increases from 33% for a current density equal to 18 mA/cm² in the case of silicon type P (100), up to 63% for j = 70 mA/cm².

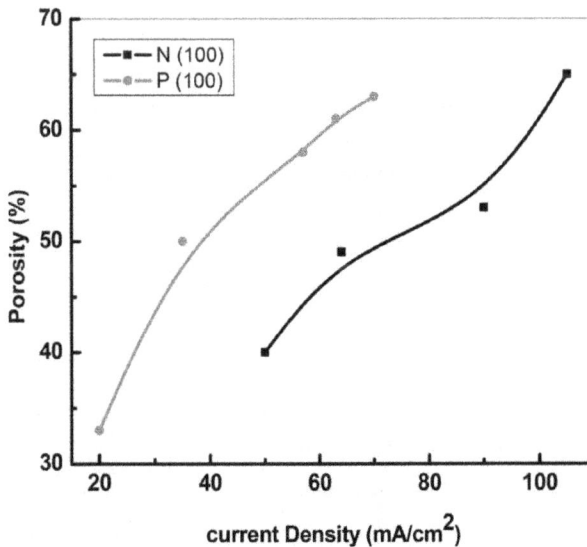

Figure 8. PS layers' porosity variation as a function of current density in (a) N-type Si (100) and (b) P-type Si (100) wafers. The used conditions are [HF] = 16% and current density of 35 mA/cm².

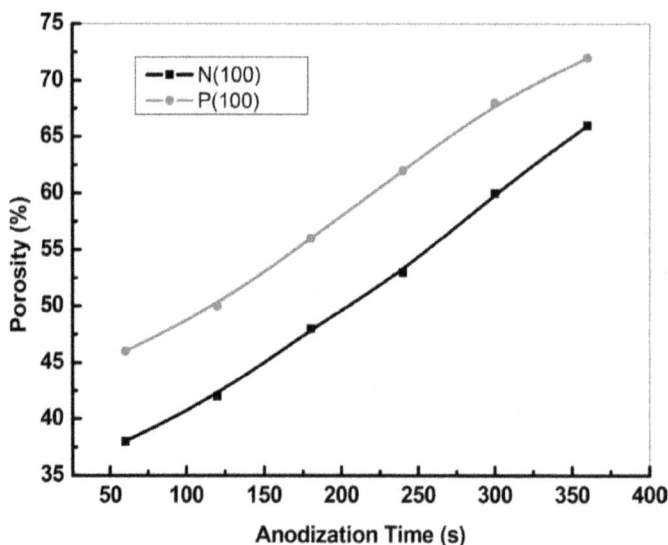

Figure 9. Porosity variation as a function anodization time measured in both Si substrates. The used conditions are [HF] = 16% and current density of 35 mA/cm².

For the N-type substrate (100), we obtained a variable porosity between 40 and 65% for a current density between 50 and 106 mA/cm².

7.2. Influence of anodization time on porosity

To know the influence of the anodization time on the porosity, we fixed the current density at j = 35 mA/cm². The concentration of [HF] = 16% in the electrolyte, and we varied the anodizing time. We obtained the curves of **Figure 9**.

In 6.1, we concluded that the thickness of the porous layer varies linearly with anodization time; from **Figure 9**, we note that the porosity also varies almost linearly over time in a function of the anodizing time. It ranges from 38 to 64% for a sample type N silicon (100), for anodizing time from 60 to 360 s, respectively. This porosity increases from 45% up to 71% for an anodizing time from 60 to 360 s for the P-type silicon sample (100).

8. Reflectivity measurement

We measured the reflection factor of two samples obtained after electrochemical anodization of N (100) and P (100) silicon substrates. The formation conditions as well as the thicknesses and porosities for the two samples are summarized in **Table 3**.

The obtained spectra are represented by **Figure 10**.

PS layers are generally used as antireflective layer in front of solar cells to reduce the light losses. The total reflectance was measured within the 400–1100 nm wavelength range

Sample	[HF] %	j (mA/cm²)	t (s)	d (nm)	P %
N (100)	16	54	360	3700	68.75
P (100)	16	70	120	2660	55

Table 3. Anodization conditions and physical parameters of Si samples used for reflectance measurement.

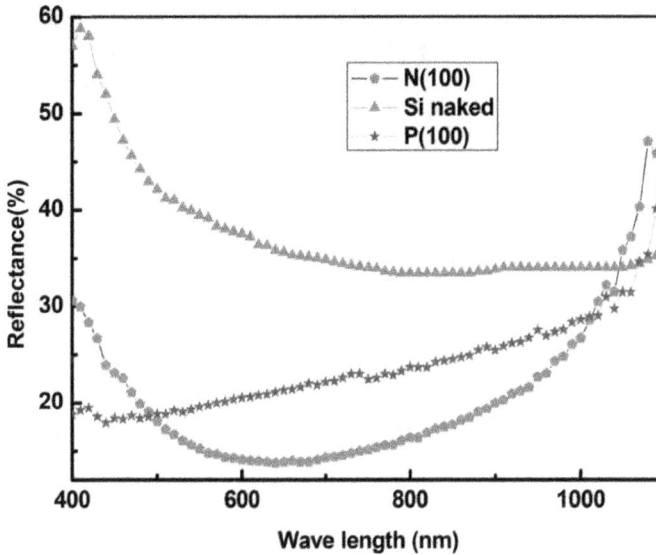

Figure 10. Variation of reflection coefficient as a function of the incident light wavelength measured in naked Si sample used as reference and N-type and P-type Si with PS layer.

with an integrating sphere, as described in **Figure 10**, where we have reported the reflection coefficient of the three studied samples: the reference Si naked (N-type) that does not undergo any chemical etching and N-type and P-type Si with porous layer. The conditions used here, such as thickness and relative porosities to the two Si samples, are summarized in **Table 1**. From **Figures 1–10**, one can deduce that the formation of PS layer reduces drastically the sample reflection coefficient due to the light entrapment in the formed pores. The reduction of losses by reflection caused by porous silicon layers indicates the antireflective character of this film type. This result agrees with what is returned in the literature [6, 8].

The antireflective activity is more significant in the N-type Si > as example for the 600 nm wavelength, the measured reflection coefficients are, respectively, 13, 20, and 37% for N-type, P-type, and naked Si substrates.

This is due to the fact that in the N type, pores have a smaller size and are uniformly distributed on the sample surface in contrary to P-type Si, where the pores are larger and spatially localized in tranches.

9. Conclusion

In the present work, we have investigated the influence of anodization time and current density upon the PS layer formation on N-type and P-type Si substrate. The formed pores in N-type Si are uniformly distributed over the sample surface with a small size, indicating an isotropic and homogenous attack of HF. However, in the case of P type, the HF etching is anisotropic; it causes the formation of large tranches composed with pores at the bottom. We have noticed that the PS layer thickness and its porosity vary linearly with the anodization time and current density. Finally, due to the pore size and distribution, we found that PS layer formed on N-type Si exhibits better antireflective activity than P-type Si, making the obtained layers important tools in the solar cell.

Author details

Salah Rahmouni[1,2*] and Lilia Zighed[3]

*Address all correspondence to: rahmouni.eln@gmail.com

1 Normal High School of Technological Education (ENSET), Skikda, Algeria

2 Department of Electrical Engineering, University of 20 August 1955, Skikda, Algeria

3 Laboratory of Chemical Engineering and Environment, University of 20 August 1955, Skikda, Algeria

References

[1] Lipinski M, Panek P, Swiatek Z, Beltowska E, Ciach RS. Double porous silicon layer on multi-crystalline Si for photovoltaic application. Solar Energy Materials & Solar Cells. 2002;**72**:271-276. DOI: https://doi.org/10.1016/S0927-0248(01)00174-X

[2] Uhlir A. Electrolytic shaping of germanium and silicon. Bell System Technical Journal. 1956;**35**:333. DOI: 10.1002/j.1538-7305.1956.tb02385.x

[3] Menna P, Di Francia G, Ferrara VLA. Porous silicon in solar cells: A review and a description of its application as an AR coating. Solar Energy Materials & Solar Cells. 1995;**37**:13-24. DOI: https://doi.org/10.1016/0927-0248(94)00193-6

[4] Lee MK, Wang YH, Chu CH. Characterization of porous silicon photovoltaic devices through rapid thermal oxidation, rapid thermal annealing and HF-dipping processes. Solar Energy Materials & Solar Cells. 1999;**59**:59-64. DOI: https://doi.org/10.1016/S0927-0248(99)00031-8

[5] Strehlke S, Bastide S, Lévy Clement C. Optimization of porous silicon reflectance for silicon photovoltaic cells. Solar Energy Materials & Solar Cells. 1999;**58**:399. DOI: https://doi.org/10.1016/S0927-0248(99)00016-1

[6] Strehlke S, Sarti D, Krotkus A, Grigoras K, Lévy-Clément C. The porous silicon emitter concept applied to multicrystalline silicon solar cells. Thin Solid Films. 1997;**297**:291-295. DOI: https://doi.org/10.1016/S0040-6090(96)09368-6

[7] Bergmenn RB, Rinke TJ, Wagner TTA, Warner JH. Solar Energy Materials & Solar Cells. 2001;**65**:355

[8] Canham LT. Silicon quantum wire array fabrication by electrochemical and chemical dissolution of wafers. Applied Physics. 1990;**57**:1046. DOI: https://doi.org/10.1063/1.103561

[9] Asoh H, Arai F, Ono S. Effect of noble metal catalyst species on the morphology of macroporous silicon formed by metal-assisted chemical etching. Electrochim Acta. 2009;**54**:5142-5148. DOI: https://doi.org/10.1016/j.electacta.2009.01.050

[10] Harraz FA, Salem AM, Mohamed BA, Kandil A, Ibrahim IA. Electrochemically deposited cobalt/platinum (Co/Pt) film into porous silicon: Structural investigation and magnetic properties. Applied Surface Science. 2013;**264**:391-398. DOI: 10.1016/j.apsusc.2012.10.032

[11] Bessais B, Ben Younes O, Ezzaouia H, Mliki N, Boujimil MF, Oueslati M, Bennaceur R. Morphological changes in porous silicon nanostructures: non-conventional photoluminescence shifts and correlation with optical absorption. Journal of Luminescence. 2000;**90**:101-109. DOI: https://doi.org/10.1016/S0022-2313(99)00617-1

[12] Halimaoui A. Determination of the specific surface area of porous silicon from its etch rate in HF solutions. Surface Science Letters. 1994;**306**:L550-L554

[13] Halimaoui A. Influence of wettability on anodic bias induced electroluminescence in porous silicon. Applied Physics Letters. 1993;**63**:1264-1266. DOI: https://doi.org/10.1063/1.109752

[14] Tsybeskov L, Fauchet PM. Correlation between photoluminescence and surface species in porous silicon: Low-temperature annealing. Applied Physics Letters. 1994;**64**:1983. DOI: https://doi.org/10.1063/1.111714

[15] Teschke O, Galembeck F, Gonçalves MC, Davanzo CU. Photoluminescence spectrum redshifting of porous silicon by a polymeric carbon layer. Applied Letters. 1994;**64**:3590. DOI: https://doi.org/10.1063/1.111207

[16] Ono H, Gomyu H, Morosalci H, Nozaki S, Shou Y, Shimazaki M, Iwaze M, Izumi T. Effects of anodization temperature on photoluminescence from porous silicon. Journal of the Electrochemical Society. 1993;**140**(12):L180-L182. DOI: 10.1149/1.2221158

[17] Koyama H, Koshida N. Photo-assisted tuning of luminescence from porous silicon. Journal of Applied Physics. 1993;**74**:6365. DOI: https://doi.org/10.1063/1.355160

[18] Lévy-Clément C, Lagoubiet A, Tomkiewicz M. Morphology of porous n-type silicon obtained by photoelectrochemical etching I. Correlations with material and etching parameters. Journal of the Electrochemical Society. 1994;**141**:958. DOI: 10.1149/1.2054865

[19] Cullis AG, Canham LT, Dosser OD. The structure of porous silicon revealed by electron microscopy. Materials Research Society Symposium Proceedings. 1992:256. DOI: https://doi.org/10.1557/PROC-256-7

[20] Levy-Clement C. Porous silicon sci, Technol, les éditions de physiques springer; 1994. pp. 327-344

[21] Chuang SF, Collins SD, Smith RL. Preferential propagation of pores during the formation of porous silicon: A transmission electron microscopy study. Applied Physics Letters. 1989;**55**(7):675-677. DOI: https://doi.org/10.1063/1.101819

[22] Vial JC, Derrien J, editors. Porous silicon in science and technology Winter. Berlin: School Les Houches, Springer-Verlag; 1994 Les Editions de Physique, Les Ulis

[23] Fathauer RW, George T, Ksendzov A, Vasquez RR. Visible luminescence from silicon wafers subjected to stain etches. Applied Physics Letters. 1992;**60**:995. DOI: 10.1063/1.106485

[24] Levy Clement. Characteristics of porous n-type silicon obtained by photo élctrochemical etching dans porous silicon science and Technology. Les éditions de physique springer; 1994. p. 329

[25] Ozanam H, CHazalviel J-N. In-situ infrared characterization of the electrochemical dissolution of silicon in a fluoride électrolyte. Journal of Electron Spectroscopy and Related Phenomena. 1993;**64/65**:395-402

[26] Lerondel G. Propagation de la lumière dans le silicium poreux application photonique. P (9-13-15) Thèse de doctorat université joseph Fourier –Gronoble I; 1997

[27] Roussel P. Micro capteur de conductivité thermique sur caissons épais de silicium poreux pour la mesure de la microcirculation sanguine. Thèse Institut national des sciences Appliquées de Lyon; 1999

[28] Halimaoui A. Porous silicon material processing properties and Technology Les éditions de physique. Vol. 1. Springer; 1994. pp. 33-52

[29] Setzu S. Réalisation et étude de structure a modulation d'indice optique en silicium poreux, Thèse de doctorat Gronoble I; 1999

Chapter 5

Digital Optical Switches with a Silicon-on-Insulator Waveguide Corner

DeGui Sun

Additional information is available at the end of the chapter

http://dx.doi.org/10.5772/intechopen.76584

Abstract

In this chapter, the quantum process of the Goos-Hänchen (GH) spatial shift is first derived out, then the coherence between spatial and angular shifts in the GH effect in the quantum state is discovered and a function of digital optical switch is developed. It is found that a waveguide corner structure always makes the reflected guide-mode have both the GH spatial and angular shifts when the incident beam is in the vicinity area of critical and Brewster angles. Meanwhile, these two GH shifts have the interesting coherent distributions with the incident angle, and only in the common linear response area the two GH shifts are mutual enhancing, then a mini refractive index modulation (RIM) of guided mode at the reflecting interface can create a great stable jump of reflected beam displacement at an eigenstate. As a result, on 220 nm silicon-on-insulator (SOI) waveguide platform, with a tapered multimode interference (MMI) waveguide a $5.0 \times 10^{18} \ cm^{-3}$ concentration variation of free carriers can cause a digital total 8–25 μm displacement of the reflected beam on the MMI output end, leading to a $1 \times N$ scale digital optical switching function. As a series of verifications, the numerical calculations, finite difference time domain (FDTD) simulations and experiments, are sustainable to the quantum GH shifts.

Keywords: silicon photonics, silicon-on-insulator, waveguide corner, coherent Goos-Hänchen spatial and angular shifts, eigenstate of total GH displacement, and digital optical switch

1. Introduction

In 1947 a novel spatial shift phenomenon of the reflected beam at an interface of two media was discovered by Goos and Hänchen, which was later referred to as the Goos-Hänchen (GH) effect [1]. Then, in 1948 it was theoretically modeled by V. K. Artmann by providing an

Artmann equation [2]. But, at that time no special attention was ever paid to this interesting phenomenon until 1972 when Chiu and Quinn further testified the simple delay process of optical scattering caused by this optical GH effect because Chiu and Quinn's work exactly presented the two polarization-determined parallel reflected beams in both theory and experiment [3]. In the past decades, based on the different media and technologies a variety of GH shift phenomena were studied in which their essential applications have been shown out [4–8]. At the condition of plane waves, the first impressive research in theory was done by Wild and Giles in [4]. Later, in 1986 Lai's team performed the more detailed theoretical investigation for the GH effect with Gaussian beam and obtained the more sustainable results, and furthermore in 2002 they extended their research to the more regular case—the complex expression for both the positive and negative electromagnetic materials at the reflection beam side and the frustrated total internal reflection (FTIR) was defined [5, 6]. These typical theoretical achievements in both plane and Gaussian waves have built a powerful fundamental for research on the GH behaviors of guided modes in a waveguide corner. Thus, in the past years, the establishments in the following three interesting topics relating to the GH effect have stimulated intensive research on new theories, materials and functionalities: (i) the integrated optical technologies due to a broad landscape of applications; (ii) the great spatial and angular shifts of Gaussian and quasi-Gaussian beams; and (iii) the control of GH effect in a micro-cavity by using the coherent light into the atomic energy levels [7–11]. With the special waveguide structures and corner-mirror materials, some novel phenomena of GH effect and the other involved effect such as Imbert-Fedorov shifts were also investigated and established [12–14].

In fact, the quantum mechanism supporting to theoretical research in the spatial shift of GH effect was established by Steinberg and Chiao in [15]. Then, in 2013 we theoretically demonstrated this quantum effect of the GH spatial shift with the consistent solutions of Maxwell equation and Schrödinger equation, and then proposed a digital electro-optic (EO) switching regime on semiconductor platform where a new metal-oxide-semiconductor (MOS)-capacitor type EO modulation method is also proposed and analyzed to realize an effective free-carrier dispersion (FCD) effect based refractive index modulation (RIM) [16]. Meanwhile, as a typical EO modulation scheme of semiconductor material, the FCD-based RIM of silicon-on-insulator (SOI) waveguide has been attracting the broad and intensive research attentions in the field of photonic integrated circuit (PIC) components and applications in the past decade [17, 18]. The similar part of all these FCD-RIM based switching schemes is that the FCD effect is imposed to a section of waveguide channel to cause an optical phase change, so their common intrinsic drawback is the FCD-induced optical absorption. As a result, there have not been any real accomplishments of product development reported so far. Consequently, the digital operations for the high-speed optical switching devices with advanced FCD-RIM schemes and device structures are expected for high-speed PIC devices [16–19].

In this chapter, by starting with the eigenstates of guided modes we first investigate the correspondences between the Fresnel equation and Schrödinger equation of guided modes and the coherence of the spatial shift and angular shift of the reflected guided modes under the GH effect. Then, the quantum processes responding to an incident angle with the eigenstates of guided modes of the MMI waveguide are discussed. Furthermore, the total displacement versus the concentration change of free-carrier holes is analyzed to discuss the performance of FCD-RIM based digital optical switch within the SOI-CMOS submicron waveguide corner structure. Furthermore, the sustainable numerical calculations, professional software simulations and experiments are discussed. Finally, the conclusions are given.

2. Spatial and angular shifts in Goos-Hänchen (GH) effect

2.1. Coherence between the spatial and angular shifts in GH effect

Figure 1 shows the schematic relationship of incident, reflected and transmitted beams across the interface of two optical media with the refractive indices n_1 and n_2, respectively. A common definition of GH shift is the effective spatial shift Δ of the reflected beam axis at the direction perpendicular to the propagating direction of the reflected beam with respect to the ideal position, which can be forwarded to the expression of the displacement S_p along the reflective interface with the Artmann equation form as [2, 5, 15].

$$S_p = -\frac{d\phi}{k_1 \cos\theta d\theta} \qquad (1)$$

where n_1 and n_2 stand for the refractive indices of the media at the reflected and transmitted sides, respectively, ϕ is the phase of the reflection coefficient ($r = R \cdot \exp(i\phi)$) if R indicates the amplitude coefficient), k_1 is the wave number ($k_1 = 2\pi n_1/\lambda_0$) if λ_0 is the optical wavelength in air, and θ and θ_t are the incident and transmitted angles of optical beam, respectively. In addition, an angular shift Θ is also caused under this GH effect.

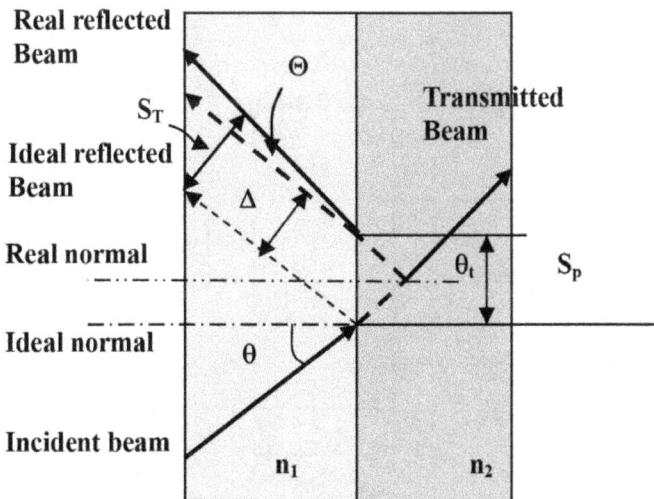

Figure 1. Schematic relationship of the incident, the reflected and the transmitted light waves at the interface of two optical media and the concepts of both GH spatial and angular shifts.

Eq. (1) tells that the phase change of reflected beam is a key parameter in the GH spatial shift, and this GH spatial shift must involve with a time delay no matter it is the light-wave traveling process of classical physics or the quantum mechanics. As demonstrated below, the optical phase change with incident angle is theoretically demonstrated to be either a quantum photonic process or the classical lightwave traveling mechanism, which is controlled by the delay time of the photon beam tunneling process in penetration depth [15, 16]. We know, in quantum mechanics theory, an electron beam in semiconductors can have a quantum-mechanical electron-wave transportation process to a potential barrier, so similarly a photon beam can also have a quantum-mechanical light-wave transportation process to a photon tunneling with a quantum effect. In 1990s, Steinberg and Chiao of the UC Berkeley did an experiment of passing through and blocking back of photon particles by using an optical system of two prisms with an air gas, then they found that the delay time of transmitted beam is shorter than the value calculated by the ray optical trajectory method, which is also referred to as an FTIR phenomenon. The FTIR phenomenon produces a quantum change of optical phase and was equivalently depicted as in **Figure 2**, where θ, θ_r and θ_t are the incident, reflected and transmitted angles of photon beam. If ϕ_d and τ_d are the phase change and the delay time of transmitted beam, respectively, after passing the penetration region of d, the transverse displacement Δz is expressed by [15].

$$\Delta z = -\frac{c}{n_1 \, \omega \cos\theta} \cdot \left[\frac{\partial \phi_d}{\partial \theta}\right]_\omega \tag{2}$$

where $\omega = 2\pi c/\lambda_0$ is the angular frequency of light-wave.

If we use ϕ replace ϕ_d in Eq. (2) to express the phase change in the penetration region of d, we immediately have solution as: $\Delta z = S_p$. We know, Eq. (2) is derived based on the electromagnetic formulism of quantum-mechanical electron-wave propagation from the Schrodinger

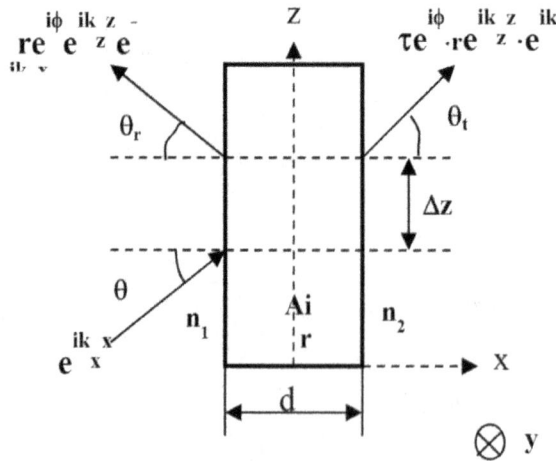

Figure 2. Schematic distribution of the one-dimensional phase shift due to the transverse displacement Δz when a photon beam tunnels a barrier with d thickness.

equation to define the transverse displacement Δz of both transmitted and reflected waves but it is the completely equivalent to the GH shift s_p defined by Eq. (1) that is evolved from the Maxwell equation of photon-wave. Thus, it turns out that the definition of the GH shift used in this work is very believable.

From Ref. [15], as shown in **Figure 2**, we have the delay time of transmitted beam during producing the shift Δz at the vertical direction as

$$\tau_d = \left[\frac{\partial \phi_d}{\partial \theta}\right]_\omega + \frac{n_2}{c}\Delta z \sin\theta \tag{3}$$

Then, by comparing **Figure 2** with **Figure 1**, the delay time in Eq. (3) can be understood as the phase change of the reflected beam after it suffers from a GH shift. Then, setting $\Delta z/c = \tau_d$ yields

$$\frac{\Delta z}{c} = \left[\frac{\partial \phi_d}{\partial \theta}\right]_\omega + \frac{\Delta z}{c} n_2 \sin\theta \tag{4}$$

Substituting Eq. (2) into Eq. (4) yields

$$-\frac{1}{n_1 \omega \cos\theta} \cdot \left[\frac{\partial \phi_d}{\partial \theta}\right]_\omega = \left[\frac{\partial \phi_d}{\partial \theta}\right]_\omega - \frac{n_2 \sin\theta}{n_1 \omega \cos\theta} \cdot \left[\frac{\partial \phi_d}{\partial \theta}\right]_\omega \tag{5}$$

Then, we obtain

$$\left[\frac{1 + n_1 \omega\cos\theta - n_2 \sin\theta}{n_1 \omega\cos\theta}\right] \cdot \left[\frac{\partial \phi_d}{\partial \theta}\right]_\omega = 0 \tag{6}$$

We know θ is an angle close to the Brewster angle or the critical angle of the optical TIR system shown in **Figure 1** and $n_1 \omega \cos\theta \gg n_2 \sin\theta$, so we have the final solutions as

$$\frac{1 + n_1 \omega\cos\theta - n_2 \sin\theta}{n_1 \omega\cos\theta} \neq 0 \text{ and } \left[\frac{\partial \phi_d}{\partial \theta}\right]_\omega = 0 \tag{7}$$

Thus, the term $\left[\frac{\partial \phi_d}{\partial \theta}\right]_\omega$ must be the critical values of the function ϕ_d with respect to θ, i.e., $\phi_d(\theta)$.

2.2. Duality of spatial and angular shifts in GH effect

In 1980s the scientists of the United States predicted, it is reasonable for the reflected beam to display a small angular deviation from the law of specular reflection [20, 21]. Until two decades later, in 2006 and 2009, the experimental results of the angular shift under the GH effect were observed in microwave and optical domains, respectively [9, 22]. In the optical experiments, the general role of GH spatial and angular shifts is modeled with both the phase and amplitude of reflected beam, then with **Figure 1**, by setting a multimode interference (MMI) with l_{mm} length, the total beam displacement is defined by a combination of the spatial shift Δ and angular shift Θ as [22]

$$S_T = \frac{1}{\cos\theta}(\Delta + l_{mm}\,\Theta) \tag{8}$$

where l_{mm} stands for the length of tapered reflective MMI waveguide. From the reflection coefficient $r = R \cdot \exp(i\phi)$, the general algebra expression of the GH shift can be expressed as

$$D = \frac{\partial \ln r}{\partial \theta} = \frac{1}{R}\frac{\partial R}{\partial \theta} + i\frac{\partial \varphi}{\partial \theta} \tag{9}$$

Then, with the propagation constant of the input guided mode β_{in}, the combination of Eqs. (8) and (9) yields the spatial and angular shifts in the GH effect as

$$\Delta = \frac{1}{\beta_{in}}\frac{\partial \varphi}{\partial \theta} = \mathrm{Im}(\ln r) \tag{10a}$$

$$\Theta = \frac{2}{(\beta_{in}^2\, w_0^2)}\frac{\partial R}{\partial \theta} = \mathrm{Re}(\ln r) \tag{10b}$$

where w_0 stands for the waist of an approximated Gaussian beam of reflected guided mode. Therefore, in terms of the set of Eqs. (10a) and (10b), the GH spatial and angular shifts involve with the phase and amplitude of the reflected beam, respectively.

For the TE- and TM-mode, the amplitude reflection coefficients R_{TE} and R_{TM} are, respectively, defined by [6, 16]:

$$R_{TE} = \frac{\cos\theta - (\eta^2 - \sin^2\theta)^{1/2}}{\cos\theta + (\eta^2 - \sin^2\theta)^{1/2}} \tag{11a}$$

$$R_{TM} = \frac{\eta^2\cos\theta - (\eta^2 - \sin^2\theta)^{1/2}}{\eta^2\cos\theta + (\eta^2 - \sin^2\theta)^{1/2}} \tag{11b}$$

where $\eta = N_{eff}/n_m$ when N_{eff} is the effective refractive index of the guide mode of input waveguide and n_m is the refractive index of corner mirror material. In Ref. [16], for the eigenstate of reflected mode on the GH spatial shift, the partial derivative meets $\partial\phi/\partial\theta = 0$, so the incident angle θ must have its corresponding eigenvalue, which creates a critical value of wave function $\varphi(x, y, z)$. From Eqs. (10b) and (11) we find that the GH angular shift Θ is dependent of the effective index N_{eff}, while N_{eff} is determined by the input waveguide material.

2.3. Understanding to the quantum spatial and angular shifts in GH effect

We know all the guided modes of an MMI waveguide are the eigenstates of quantum physical process and the corresponding effective indices are the eigenvalues of all the refractive indices of optical beam at the phase velocities [23, 24]. So, if we set the input waveguide channel as a single-mode, N_{eff} would be only the eigenvalue of the single-mode, then it finally determines one of the GH angular shifts. Thus, for a waveguide structure, if $\varepsilon_r(x)$ stands for the relative dielectric constant of waveguide material, the two-dimensional (2D) Maxwell wave equation of the propagation constant β and light wave function $\varphi_{(x, z)}$ for the TE- and TM-mode of optical waveguide are expressed by (12a) and (12b), respectively, as [25].

$$2j\beta \frac{\partial\varphi(x, z)}{\partial z} = \frac{\partial^2 \varphi(x, z)}{\partial x^2} + k_0^2(\varepsilon_r^2 - N_{eff}^2)\varphi(x, z) \tag{12a}$$

$$2j\beta \frac{\partial\varphi(x, z)}{\partial z} = \varepsilon_r \frac{1}{\partial x}\left(\frac{1}{\varepsilon_r} \frac{\partial\varphi(x, z)}{\partial x}\right) + k_0^2(\varepsilon_r^2 - N_{eff}^2)\varphi(x, z) \tag{12b}$$

For a quantum system, if $U_{(x)}$ and $m_{(x)}$ stand for the arbitrary potential and the effective mass, respectively, with the effective mass approximation and the Plank constant $\hbar$, the time-dependent relation for the photon wave function $\Psi_{(x, t)}$ can be expressed by a Schrödinger equation as

$$j\hbar \frac{\partial\Psi(x, t)}{\partial t} = \frac{\hbar^2}{2} \frac{\partial}{\partial x}\left(\frac{1}{m(x)} \frac{\partial\Psi(x, t)}{\partial x}\right) + U(x)\Psi(x, t) \tag{13}$$

Comparing (13) with (12a) and (12b) yields the conclusions that the TE-mode corresponds to the case that the effective mass is independent of x, while the TM-mode corresponds to the case that the effective mass is dependent of x. Hence, we can find the following correspondences for qth mode:

$$z \leftrightarrow t, \tag{14a}$$

$$2j\beta \leftrightarrow j\hbar, \tag{14b}$$

$$\frac{\partial^2}{\partial x^2} \leftrightarrow \frac{\hbar^2}{2m} \frac{\partial^2}{\partial x^2} \text{ for TE-mode} \tag{15a}$$

$$\varepsilon_r \frac{1}{\partial x}\left(\frac{1}{\varepsilon_r} \frac{\partial}{\partial x}\right) \leftrightarrow \frac{\hbar^2}{2} \frac{\partial}{\partial x}\left(\frac{1}{m(x)} \frac{\partial\Psi(x, t)}{\partial x}\right) \text{ for TM-mode} \tag{15b}$$

$$k_0^2(\varepsilon_r^2 - N_{eff}^2)\varphi(x, z) \leftrightarrow U(x)\Psi(x, t) \tag{16}$$

It turns out that the angular shift defined by Eq. (10b) is a quantum selection from multiple eigenstates of the reflection angle under the GH effect. It turns out from relation (14a) that the distance z of light wave traveling corresponds to the time t of photon beam tunneling, while in quantum mechanism, there is no variation of particle velocity, so this corresponding relation is very reasonable. It turns out from relation (14b) again that the propagation constant relates the effective wave number of the guided mode in the traveling-through channel, while the Plank constant $\hbar$ relates the eigenstate of particle with its mass in tunneling barrier, so the guided mode is limited to a quantum state. Therefore, the relations, (15a), (15b) and (16) are used to analyze the field-distributions of guided mode at its eigenstate.

3. A digital optical switch principle with the waveguide corner

3.1. Device concept and condition for the digital optical switch

With the SOI-based waveguide corner mirror (WCM) structure comprising semiconductor (silicon) waveguide channel material, oxide (SiO_2 or SiON) corner mirror material and metal material for electrodes, a new regime of 1xN scale digital optical switches was proposed as shown in **Figure 3**. This special regime of digital optical switches composed of a single-mode input waveguide, a tapered multi-mode corner waveguide, a corner mirror and N single-mode output waveguides is proposed, where N is an integer. In this device regime, the single-mode output waveguide channels are all in conjunction with the output end of the tapered MMI corner waveguide [15, 16]. Then, the free-carrier concentration variation at the interface area of mirror and waveguide is controlled to realize the FCD-RIM of silicon of this WCM structure [16–19].

In optical switching operations, as shown in **Figure 3**, all the possible angles of the reflected guided mode are no longer the same due to an angular shift of the GH effect. Namely, the angular shift in this project makes the reflection angle different from the input angle and then creates a great total displacement at the output end of tapered MMI waveguide with a combined effect of the spatial shift and angular shift as defined by Eq. (8).

3.2. Discussion for the mode distributions of MMI waveguide

In this 1xN type digital EO switch shown in **Figure 3**, the tapered MMI waveguide is a critical part as it not only forms a reflecting interface with the corner mirror, but also transfers the reflected optical beam from a single-mode input waveguide to its corresponding single-mode output waveguide. Thus, at the input end this tapered MMI waveguide is required not only to have the corresponding eigenstate guided mode to match the eigenstate reflected-mode of the GH effect, but also have the length to maintain the single-mode state of reflected guided mode from the input waveguide to the corresponding output waveguide.

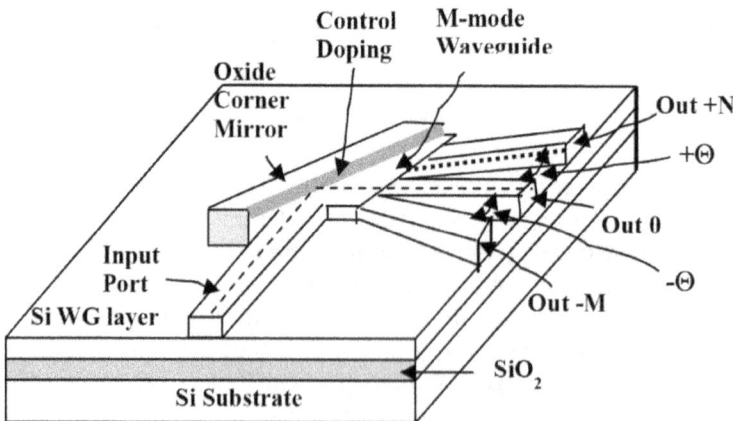

Figure 3. Schematic architecture of the SOI-WCM for the GH effect based 1x3 optical switch with CMOS-compatible sub-micron scale waveguides and FCD-RIM.

With the consideration for both the GH spatial and angular shifts in the FTIR process of WCM structure, the tapered MMI waveguide was set to meet the conditions at its input end so that it can produce a sufficient separation between two adjacent output modes at its output end. As explained above, the tapered MMI waveguide is assumed to contain N modes with the numbers $q = 0, 1, ..., N-1$, the GH angular shift defined by Eq. (10b) is a quantum selection for the eigenstate reflection angle of a reflected guide-mode to match one of the multiple guided modes of tapered MMI waveguide at its input end which has an effective index N_{eff}. **Figure 4** schematically depicts such a tapered MMI waveguide where W_i and W_0 are its widths at the input and output ends, respectively, and the total displacement s_T is determined by the spatial and angular shifts as defined by Eq. (8).

In the MMI waveguide a specified factor is the beat length of the two lowest-order modes (i.e., the two key modes) L_π, which is defined by the propagation constant difference of these two modes $\Delta\beta_{01} = \beta_0 - \beta_1$ as

$$L_\pi = \frac{\pi}{\Delta\beta_{01}} \approx \frac{4\,n_1\,W_i^2}{3\,\lambda_0} \tag{17}$$

The 1-mode length is required to be $L_{1m} = \left(\frac{3}{4}\right) L_\pi$, then we have:

$$L_{1m} = \left(\frac{3}{4}\right) L_\pi \approx \frac{n_1\,W_i^2}{\lambda_0} \tag{18}$$

For the qth mode eigenstate guided mode, the lateral wavenumber k_{xq} and the propagation constant β_q should be related to a core refractive index n_1 of waveguide as [23, 24]

$$k_{xq}^2 + \beta_q^2 = k_0^2 n_i^2 \text{ with} \tag{19a}$$

$$k_0 = 2\pi/\lambda_0, k_{kq} = (q+1)\pi/W_i \tag{19b}$$

Accordingly, the propagation constant spacing of any high order mode from the fundamental mode is expressed as

$$\Delta\beta_{0q} = \beta_0 - \beta_q \approx \frac{q(q+2)\pi}{3\,L_\pi} \tag{20}$$

3.3. Matching condition between the RIM modulation and the guided mode

In the FCD-RIM process of an SOI-waveguide device, if the concentration of free-carriers in silicon waveguide has a variation for the guided mode, at $\lambda = 1550$ nm the RIM of silicon is defined by the refined Drude-Lorentz model as [18, 19].

$$\Delta n = -\left[8.8 \times 10^{-22}\,\Delta N_e + 8.5 \times 10^{-18}\left(\Delta N_h\right)^{0.8}\right] \tag{21a}$$

$$\Delta\alpha = 8.5 \times 10^{-18}\,\Delta N_e + 6.0 \times 10^{-18}\,\Delta N_h \tag{21b}$$

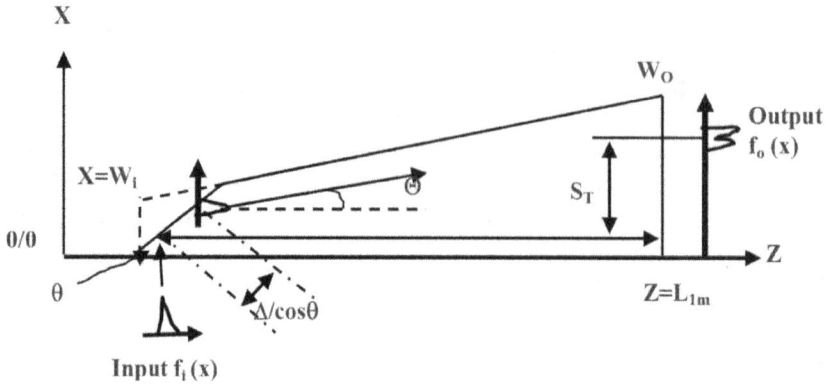

Figure 4. Schematic structure of the tapered MMI corner waveguide for transferring the reflected optical beam from the input waveguide to an output waveguide at the single-mode state with 1-mode length $Z = L_{1m}$, where the spatial/angular shifts, and the total displacement of GH effect are labeled.

where Δn and $\Delta \alpha$ are the changes of refractive index and absorptive coefficient, respectively, resulting from the variation in free carrier concentrations; ΔN_e and ΔN_h are the concentration variations of free electrons and free holes, respectively. Eqs. (21a) and (21b) imply that the FCD-RIM effect also causes an extra optical loss in its refractive index modulation. For the reflection of the guided mode at the waveguide/corner-mirror interface, the incident and reflected optical fields should be expressed as [26, 27].

$$E_i = \exp.[i(\omega t - k_i z)] \text{ and } E_r = R \cdot \exp[i(\omega t - k_r z)] \qquad (22)$$

where the reflection coefficient R is defined by Eqs. (11a) and (11b) for TE- and TM-mode, respectively. Finally, at the qth independent mode in the FTIR process from the input mode to the reflected mode, we have a transmission equation of Fresnel formula between the critical angle θ_c and the effective index N_{eq} as.

$$n_2 \approx N_{eq} \sin[\theta_c + \Theta_q] \qquad (23)$$

We know, there are more than two guided modes in the input end of an MMI waveguide and any independent guided mode has an independent N_{eq} value [23, 24], so it is the exclusive value to determine a jump of the reflected beam in this GH effect.

4. Simulations and analyses for the FCD-based modulation performance

In Section 2.2, we theoretically have proved that both the GH spatial and angular shifts form a total displacement of reflected beam at the output end of MMI waveguide. So, the performance of this digital optical switch is based on the dual GH shifts.

4.1. Analysis for both the spatial and angular shifts of GH effect

By selecting the CMOS-compatible SOI waveguide and taking the silicon layer thickness as H = 220 nm, ridge height as h = 130 nm, silicon refractive index as n_1 = 3.43, and SiO_2 refractive index as n_2 = 1.46, then for the rib widths: 350 and 400 nm, with the finite-difference time-domain (FDTD) software we obtain the effective indices and the corresponding effective widths of the guided modes at λ=1550 nm and x-polarization as depicted in **Table 1**.

As shown in **Figures 3** and **4**, the GH effect happening on the FTIR interface has the changes of two physical parameters—the position and angle of reflected beam, which both strongly depend on the incident angle θ when it is close to the Brewster critical angle θ_c. As depicted in **Table 1**, the rib widths of 350 and 400 nm have the effective indices as N_{eff} = 2.3368 and N_{eff} = 2.4114, respectively, which lead to the critical angles as θ_c = 38.7 and 37.3°, respectively. Hence, for these two rib widths with the theoretical models defined by Eq. (10a) we first display the dependences of the GH spatial shift Δ on the incident angle θ of optical guided mode as shown in **Figures 5(a)** and **5(b)**, respectively. Note from **Figure 5** that a sharp change can be caused by a mini-change of incident angle at a vicinity of the Brewster critical angle θ_c, and the GH spatial shifts are different between the two different cases of incident angle: smaller and greater than θ_c.

In our previous work, the switching scale potential of digital optical switch scheme was only based on the GH spatial shift. However, as analyzed above, in a real FTIR process the reflected beam not only has an anti-trajectory rule phenomenon of reflecting position, but it also predicts a non-specular reflective angle shift, thus as a continuous research project, these two special photonic phenomena in the FTIR process are both dependent on one quantum state of photonic tunneling process. In the same manner, when the input single-mode waveguide has the rib widths of 350 and 400 nm, with the data used for **Figure 5** and Eq. (10b) we further obtain the simulations for the dependences of the GH angular shift Θ of reflected mode on the incident angle θ as shown in **Figures 6(a)** and **6(b)**, respectively. Note from **Figures 5** and **6** that, under the quantum GH effect, both the spatial and angular shifts of reflected beam have the sharp responses to the incident angle and are mutually influenced. However, the difference between these two shifts is that the spatial shift Δ is plus while the angular shift Θ is minus, which can be manipulated to realize a switching function with the WCM structure.

By selecting the rib width of 350 nm, the dual dependences of spatial and angular shifts on the incident angle is shown in **Figure 7**. Two significant interesting points should be found in **Figure 7** as that (1) in a vicinity of Brewster critical angle θ_c that is at 38.95°, the incident angle greater than θ_c must be selected where the angular shift is extremely big at the minus direction and (2) the two GH shifts have one common sharp linear response range. In **Figure 6**, there are five different areas in the whole distribution of the two GH shifts as: (1) In the area of incident angle smaller than 38.5° only a slow angular shift exists; (2) in the area of incident angle from 38.5 to 38.91° the two GH shifts are reverse; (3) in the area of incident angle from 38.91 to 38.95°, the two GH shifts have a common linear response area; (4) in the area of incident angle from 38.96 to 39.30° that is exactly greater than θ_c the two GH shifts are reverse again; and (5) in the area of incident angle greater than 39.3° only a slow spatial shift exists. In the common linear response area of these two GH shifts we find that at the incident angle of 38.91°, the spatial GH shift reaches its minimum value of 3.82 μm, while the GH angular

Rib width, W_r (nm)	350	400
Effective index at E_x, N_{eff}	2.3368	2.4114
Effective mode-width at E_x, W_{eff} (nm)	~350	~400

Table 1. Single-mode operations of two rib widths.

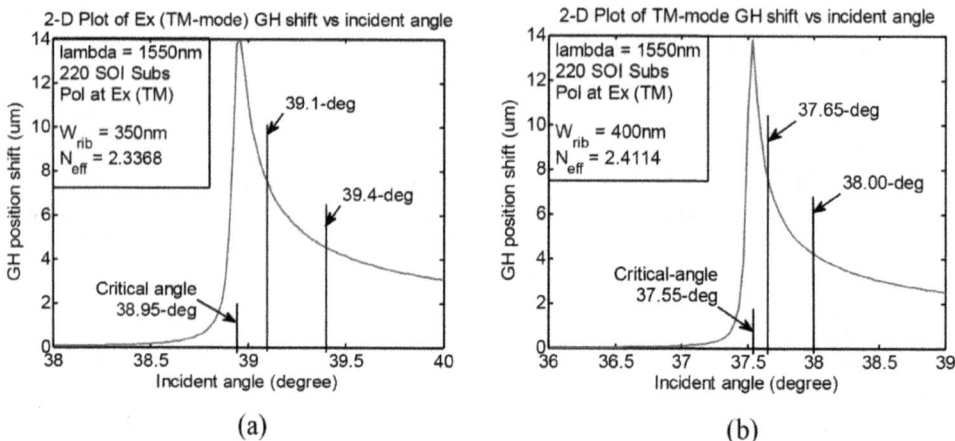

(a)

(b)

Figure 5. Incident angle dependence of the spatial shift of reflected beam under the GH effect with the WCM structure at the x-polarization: (a) and (b) are for 350 and 400 nm rib widths, respectively, where the typical values of incident angle are marked.

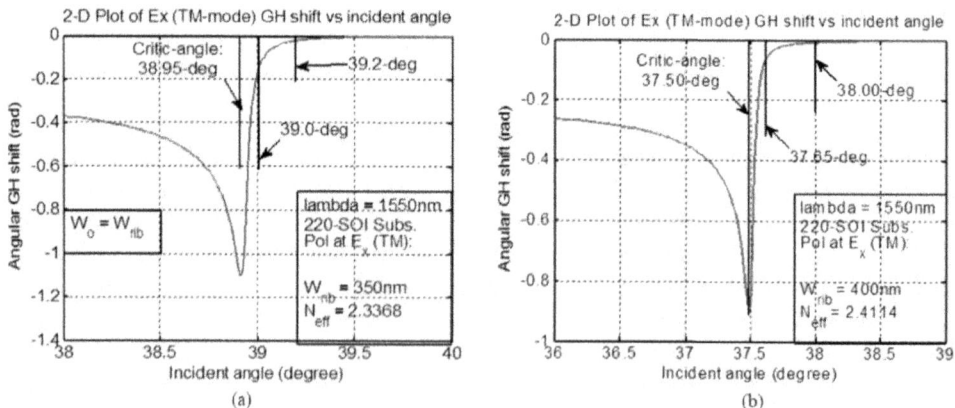

(a)

(b)

Figure 6. Incident angle dependence of angular shifts under the GH effect with the WCM structure at x-polarization: (a) and (b) are for 350 and 400 nm rib widths, respectively, where the typical values of incident angle are marked.

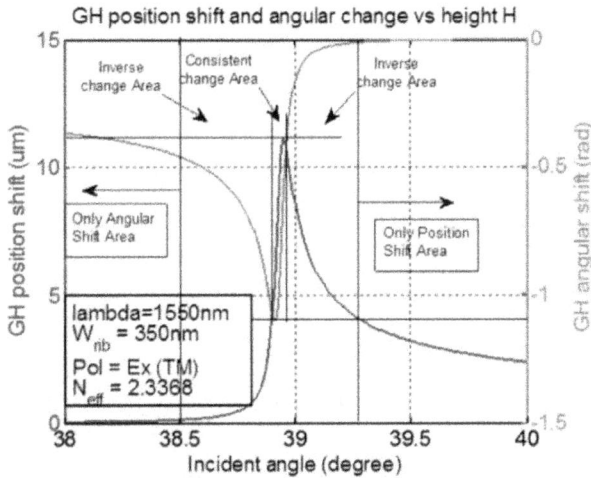

Figure 7. For the 350 nm rib width and at 1550 nm wavelength the incident angle dependences of both the spatial and angle shifts under the GH effect in which the characteristics of GH shifts are categorized.

shift reaches its maximum value of −1.10-rad. In contrary, when the incident angle is in the range of 38.94–38.95°, the GH spatial shift reaches its maximum value of 11.14 μm, while the GH angular shift reaches its lowest value of −0.35-rad. These characteristics of the common response area of GH shifts are very conducive to the designs of digital optical switches.

4.2. Analysis for the mode distribution at the input end of MMI waveguide

In order to meet the mature fabrication technique for future experiments, we select the CMOS-compatible 220 nm standard SOI-PIC platform with 130 nm rib height, at $\lambda = 1550$ nm the refractive indices for core and cladding layers are 3.43 and 1.46, respectively. For the input waveguide having a rib width of 350 nm, at the input end of MMI waveguide as shown in **Figure 8(a)**, we calculate the optical field distribution of reflected guided mode as depicted in **Figure 8(b)**. Note that the mode size is approximately 2.00 μm, so at the input end of the MMI waveguide the smallest center-center spacing between two adjacent ports is 2.00 μm, then total width should be set in the range from 2.00 to (2.00 + 11.4) μm.

4.3. Investigation for the total GH displacements at the MMI output end

In accordance with the simulations of both the spatial and angular shifts of GH effect shown in **Figures 5** and **6**, we determine the input end width $W_i = W_{eff} + \Delta$. As an illustration, by taking W_i as 2.5 μm, we calculate L_{1m} with Eq. (18) at first, and then with Eq. (8) obtain the total displacement S_T at the output end of the MMI waveguide in its linear response range to incident angle as shown in **Figure 9**.

Note from **Figure 9** that the sharp response range is the common linear response area of the two GH shifts in **Figure 6**, which is in the range of incident angle from 38.91 to 38.95° and

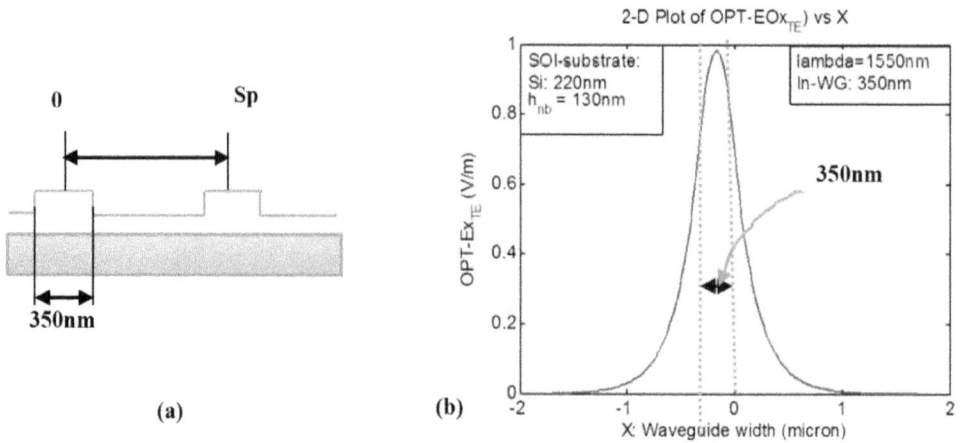

Figure 8. Optical field distribution of single-mode at the input end of MMI waveguide channel having a ridge width of 350 nm: (a) the input port cross-sectional view and (b) the optical field distribution.

labeled as a consistent change area, and in the other two incident angle areas, namely, it is smaller than 38.91° and greater than 38.96°, there are two the moderate response ranges of s_T to the incident angle. Such the distribution characteristics of the total displacement s_T at the output end of MMI waveguide are very conducive to the optimization of digital optical switch performance. Thus, it turns out that **Figure 9** is paramount important for investigating the maximum feasible switching scale and designing the best optical switch with the WCM structure and the FCD-RIM.

From **Figures 5** and **6** we know the sharp-response area of GH effect to the incident angle is a consistent area between the spatial shift and the angular shift under this GH effect, and **Figure 9** proves that the total displacement is co-contributed by the spatial and angular shifts. So, as a result, in the incident angle range from 38.83 to 39.16° it presents a sharp linear response area to the incident angle for a given MMI width at its input end. Therefore, for this 350 nm rib waveguide we first select the incident angles as 38.7, 38.9 and 39.1° in the sharp-response area of GH effect in **Figure 9** that are all very close to the critical angle and then at the wavelength of 1550 nm obtain the dependence of the total displacement of the reflected beam on the free-carry hole concentration change (HCC) for the two given MMI waveguide width values at its input end: 2.5 and 5.0 μm as shown in **Figure 10(a)** and **(b)**, respectively. In this process, as described above, the HCC activates an RIM of silicon material at the reflective interface between the silicon material of waveguide and the silicon-dioxide of corner mirror as defined by the Drude-Lorenz Eq. (21). In fact, **Figure 10** presents one significant attribute of the GH-effect based total displacement of reflected beam when the HCC is changed from 0 to $2.5 \times 10^{18}\ cm^{-3}$. Note that the displacement immediately passes through an unstable quantum jump and stabilizes on the values of −15 and −30 μm for the MMI widths of 2.5 and 5.0 μm, respectively, irrespective of the HCC continuous increasing. So, the quantum switching property has been shown out, but such a complex process will take much more research before realizing a digital optical switch.

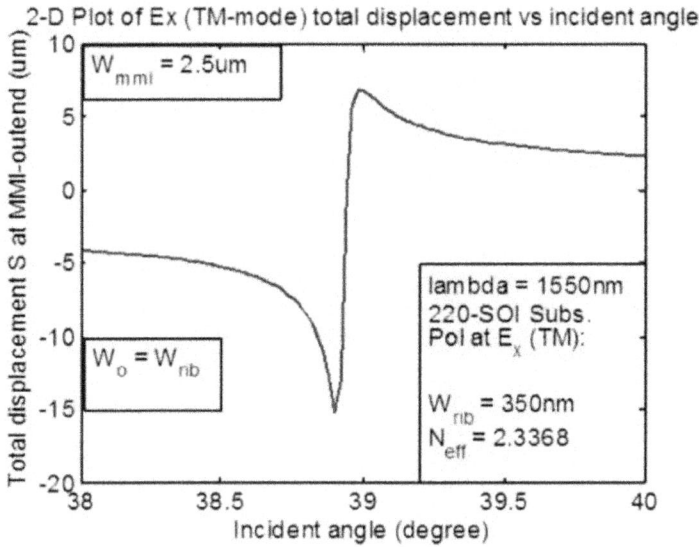

Figure 9. The total displacement at the interface of reflected beam for the rib width 350 and W_i = 2.5 μm at 1550 nm wavelength.

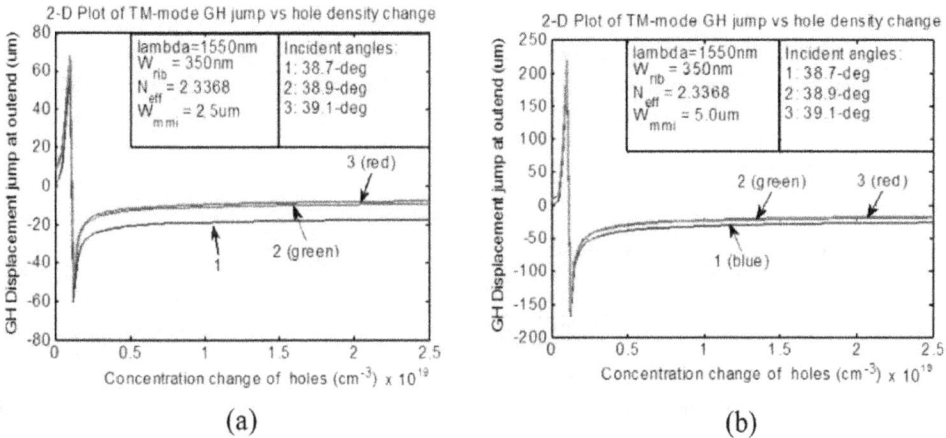

Figure 10. Dependence of the total space displacement of reflected guided mode caused by two GH shifts for the waveguide width of 350 nm at the sharp response area with respect to three different incident angles: (a) and (b) for the input end width of MMI waveguide: 2.5 and 5.0 μm, respectively.

Similarly, by selecting three incident angles: 39.1, 39.3 and 39.5° that are in the moderate area of **Figure 9** for the incident angle greater than the critical angle we obtain the total displacement dependence of the reflected beam on HCC as shown in **Figure 11** where (a) and (b) are still for the two MMI waveguide input-end widths: 2.5 and 5.0 μm, respectively. Note from **Figure 11** that the HCC of 5.0×10^{18} cm^{-3} can cause the total displacement to have a stable quantum

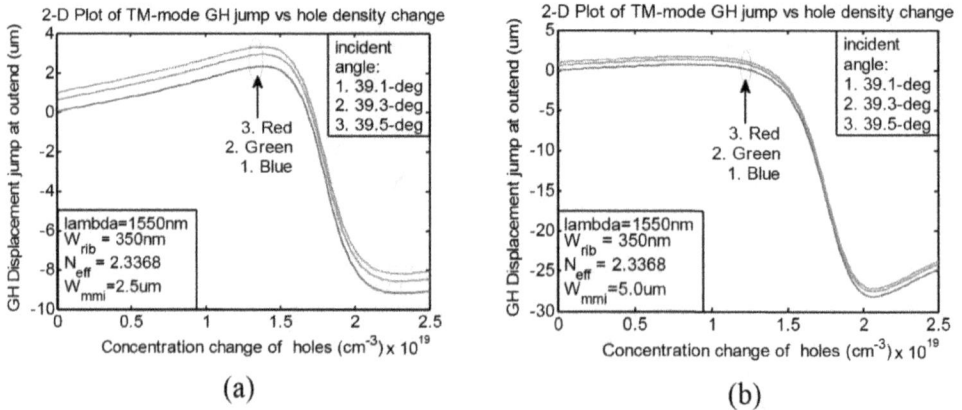

Figure 11. Dependence of the total displacement on the HCC under GH effect for waveguide width of 350 nm in the moderate response area with respect to three incident angles: (a) and (b) are for the input end widths of MMI waveguide: 2.5 and 5.0 μm, respectively.

jump, which is −8 and −25 μm for the MMI waveguide widths of 2.5 and 5.0 μm at the input end, respectively. Although the total displacements have the different jump amplitudes for these two MMI waveguide widths they need the same HCC value of $5.0 \times 10^{18} \, cm^{-3}$ to cause an FCD-RIM of about $6.0 \times 10^{18} \, cm^{-3}$ via Eq. (21). Hence, as a substantial application and one of the objectives of this work, a digital switching operation can be immediately convinced. With the free-carrier concentration control structure MOS-capacitor [16, 19], the HCC of $5.0 \times 10^{18} \, cm^{-3}$ can be controlled to be 3–5 stages by controlling both the gate and source voltages, then $1 \times N$ type digital optical switching function could be realized, where the scale metric N depends on the system design and the requirement for the isolation between any two outputs.

5. Verification for the GH shifts

5.1. Simulations for the GH spatial shift with finite difference time-domain (FDTD) software

As early as the end of 1990s, we had started to investigate the optical output performance of waveguide corner mirrors with the theoretical modeling, numerical and professional software simulations and published the achievements in 2009 [27]. Based on the numerical calculations and finite difference time-domain (FDTD) simulations for the optical power transfer efficiency of SOI waveguide corner mirrors, an SOI rib waveguide structure is designed as shown in **Figure 12(a)**, then in order to specially simulate the impact of the GH shift on the optical power transfer efficiency, a waveguide corner mirror (WCM) structure is designed to have one input single-mode waveguide channel and one output single-mode waveguide channel as shown in **Figure 12(b)**, where the reflecting mirror is created with a step fabrication of deep etching as shown in **Figure 12(c)**.

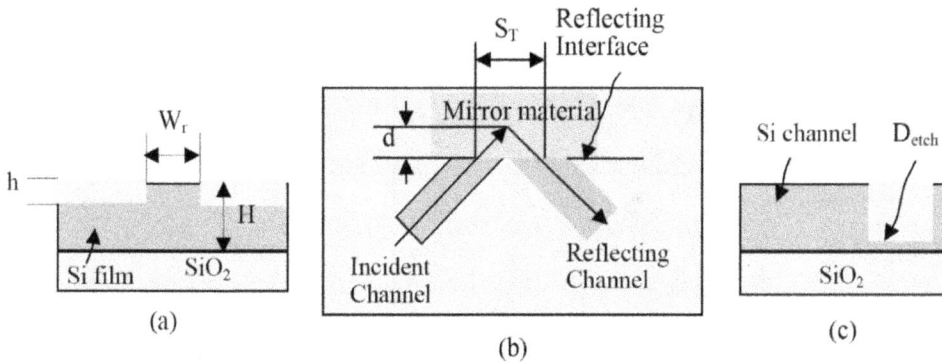

Figure 12. Schematic SOI rib waveguide corner mirror (WCM) structure: (a) the cross-sectional view of rib waveguide; (b) the distribution of waveguide channels and corner mirror with a GH shift d and (c) the deep etched part as corner.

In **Figure 12(b)** the position d is set to create a previous set GH spatial shift as: $S_T = -2d \cdot tan(\theta)$. With the feasible designs and fabrications of the real devices, two standards of SOI substrate: thick and thin silicon films are selected and the corresponding waveguide structures for each standard for single-mode operations are depicted in **Table 2**. Then, by setting the reflecting interface roughness as $\sigma = 100-500 \,\mathring{A}$ and the tilt-angle as $\varphi = 0, 1, 2°$, with the thick SOI standard and large-scale waveguide rib width of 4.0 μm, the numerical calculation as shown in **Figure 13(a)**. Note that both the position and tilt-angle of the reflecting interface play the significant impacts upon the optical power transfer efficiency. In the same manner, by selecting the roughness as $\sigma = 100 \,\mathring{A}$ and the tilt-angle of reflecting interface as $\varphi = 0, 1°$ with the thin SOI substrate listed in **Table 2**, the FDTD simulation results of the GH shift dependence of optical power transfer efficiency as shown in **Figure 13(b)**. One novel finding from **Figure 13(b)** is that the GH shift dependence curves of the optical power transfer efficiency of TE- and TM-mode have an intersection, which could never happen to any parameter dependence curves, so this novel phenomenon implies the discontinuous characteristics of the optical power transfer efficiency when the GH shift impacts the TIR process. In addition, the minus d value gives the highest optical power efficiency, matching the situation of the plus S_T value.

5.2. Design, fabrication and test of SOI waveguide corner mirror structure

We designed, fabricated and characterized the WCM structures in 2009-2011 and presented the achievements in 2013 [19]. Based on the fabrication condition of SOI waveguides, we

SOI substrate standard	Thick Si film, 4.0 μm	Thin Si film, 1.5 μm
Rib width, W_r (μm)	4.0	2.0
Rib height, h (μm)	1.0	0.5

Table 2. Two SOI standards and the corresponding rib widths for single-mode operations.

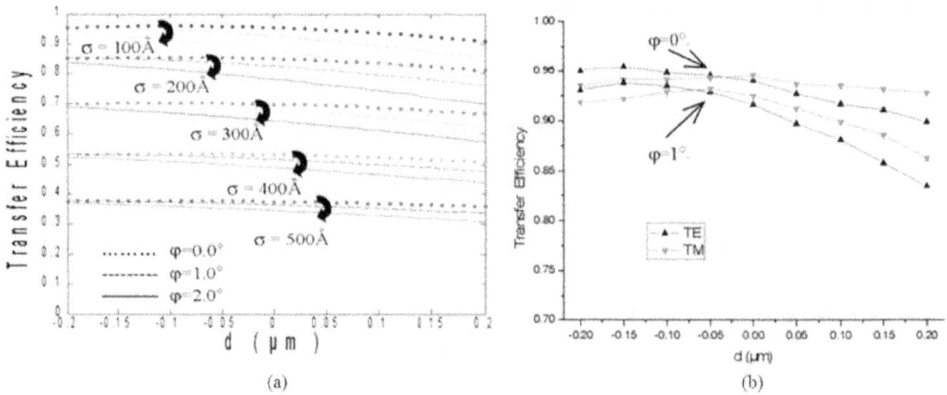

(a)   (b)

Figure 13. Dependence of optical output on the GH spatial shift GH effect for the thick SOI standard: (a) the numerical calculation results with the thick SOI substrate and (b) the FDTD simulations with the thin SOI substrate.

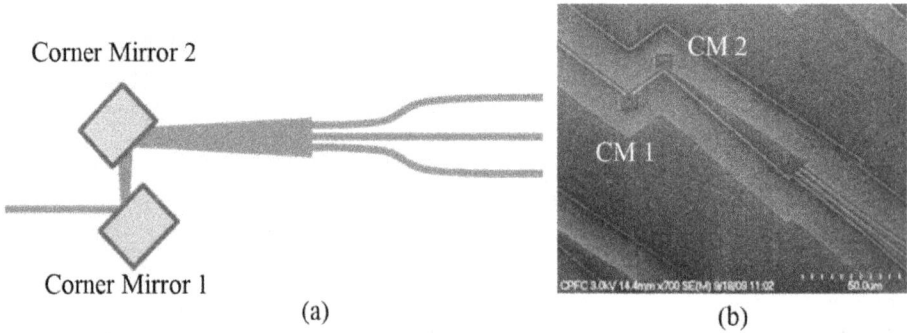

(a)   (b)

Figure 14. Design & fabrication of SOI-WCM having 1 input waveguide channel and 3 output waveguide channels: (a) the schematic layout of 1 × 3 SOI-WCM and (b) the perspective view SEM image of a fabricated device sample.

(a)   (b)   (c)   (d)   (e)

Figure 15. Testing results of optical outputs at three ports for the fabricated devices with five GH shift values that can be caused by FCD-based RIM with 2d of: (a) −0.3 μm; (b) −0.15 μm; (c) 0; (d) +0.15 μm; and (e) +0.30 μm.

selected the thin SOI substrate and then designed the 1 × 3 type WCM structure according to the regime of digital optical switches for evaluating the optical performance of both the GH shift effect and indirectly measuring the switching possibilities of digital optical switch as shown in **Figure 14(a)** and then the SEM image of fabricated device sample is shown in **Figure 14(b)**. By setting 5 values of d as: -0.15, -0.07, 0, 0.07 and 0.15 μm on each corner, we obtained 5 optical output modes as shown in **Figure 15** from (a) through (e). A very significant point is that the two minus d values: −0.15 and −0.07 μm give the obvious higher optical output at the middle port than the case of d = 0, but there is no obvious difference between these two d values, which is probably due to the quantum effect of the two GH shifts apart from the non-uniformity of fabrication. For the two plus d values: 0.15 μm, the output at the right port is higher than the cases of both d = 0 and d = 0.07 μm due to GH effect.

6. Conclusions

In this chapter, based on the coherent quantum process of the GH spatial and angular shifts, the new mechanism of substantial digital optical switches is investigated with a WCM structure. For the new regime of digital optical switches, an ideal optical refractive index modulation is FCD effect of semiconductor silicon for ultrahigh speed switching operations of pico- and nanosecond levels, but the other refractive index modulation such as the thermos-optic modulation can also be selected to realize the μs level operations. In addition, the mutual enhancing contributions of the two quantum GH shifts can be further developed to optimize the total displacement of reflected beam for digital optical switches. Therefore, this work will be helpful for research and development of high integrable and high-speed optical and photonic switches on the platform of silicon photonics.

Acknowledgements

This work is co-sponsored by the Innovative R&D Fund of CUST/China, the Research Program of Ontario Centre for Excellences (OCE) /Canada, and the in-kind invest of D&T Photonics, a startup spun-off from University of Ottawa. Author is very grateful for Prof. Trevor J. Hall for his supervising in design and fabrication of the experimental samples, and also thanks his graduated students: Xiaoqi Li and Jia Yi for their supporting works.

Author details

DeGui Sun[1,2]*

*Address all correspondence to: sundg@cust.edu.cn

1 School of Science, Changchun University of Science and Technology, Changchun, China

2 Centre for Research in Photonics, University of Ottawa, Ottawa, ON, Canada

References

[1] Goos F, Hänchen HH. Ein neuer und fundamentalater versuch zur totalreflexion. Annals of Physics. 1947;**1**:333-346. DOI: 10.1002/andp.19474360704

[2] Artmann K. Berechnung der seitenversetzung des totalreflektierten strahles. Annals of Physics. 1948;**6**:87-102. DOI: 10.1002/andp.19484370108

[3] Chiu KW, Quinn JJ. On the Goos-Hänchen effect: A simple example of a time delay scattering process. American Journal of Physics. 1972;**40**:1847-1851. DOI: 10.1119/1.1987075

[4] Wild WJ, Giles CL. Goos-Hänchen shifts from absorbing media. Physics Review. 1982;**A25**:2099-2101. DOI: 10.1103/PhysRevA.25.2099

[5] Lai HM, Chen FC, Tang WK. Goos-Hänchen effect around and off the critical angle. Journal of the Optical Society of America. 1986;**A3**:550-557. DOI: 10.1364/JOSAA.3.000550

[6] Lai HM, Chan SW. Large and negative Goos-Hänchen shift near the Brewster dip on reflection from weakly absorbing media. Optics Letters. 2002;**27**:680-682. DOI: 10.1364/OL.27.000680

[7] Qing DK, Chen G. Goos-Hänchen shifts at the interfaces between left- and right-handed media. Optics Letters. 2004;**29**:872-874. DOI: 10.1364/OL.29.000872

[8] Oh GY, Kim DG, Choi YW. The characterization of GH shifts of surface plasmon resonance in a waveguide using the FDTD method. Optics Express. 2009;**17**:20714-20720. DOI: 10.1364/OE.17.020714

[9] Merano M, Aiello A, Exter MP, Woerdman JP. Observing angular deviations in the specular reflection of a light beam. Nature Photonics. 2009;**3**:337-340. DOI: 10.1038/NPHOTON.2009.75

[10] Ziauddin SQ, Quama S, Zubairy MS. Coherent control of the Goos-Hänchen shift. Physics Review. 2010;**A81**:023821. DOI: 10.1103/PhysRevA.81.023821

[11] Wang LG, Chen H, Zhu SY. Large and negative Goos-Hänchen shift near the Brewster dip on reflection from weakly absorbing media. Optics Letters. 2005;**30**:2936-2938

[12] Aiello A. Goos-Hänchen and Imbert-Fedorov shifts: A novel perspective. New Journal of Physics. 2012;**14**:1-12. DOI: 10.1088/1367-2630/14/1/013058

[13] Bliokh KY, Aiello A. Goos-Hänchen and Imbert-Fedorov beam shifts: An overview. Journal of Optics. 2013;**15**:014001. DOI: 10.1088/2040-8978/15/1/014001

[14] Rechtsman MC, Kartashov YV, Setzpfandt F, Trompeter FH, Torner L, Pertsch T, Peschel U, Szameit A. Negative Goos-Hänchen shift in periodic media. Optics Letters. 2011;**36**:4446-4448. DOI: 10.1364/OL.36.004446

[15] Steinberg AM, Chiao RY. Tunneling delay times in one and two dimensions. Physics Review. 1994;**A49**:3283-3295. DOI: 10.1103/PhysRevA.49.3283

[16] Sun DG. A proposal for digital electro-optic switches with free-carrier dispersion effect and Goos-Hanchen shift in silicon-on-insulator waveguide corner mirror. Journal of Applied Physics. 2013;**114**:104502. DOI: 10.1063/1.4820378

[17] Campenhout JV, Greens WMJ, Vlasov YA. Design of a digital, ultra-broadband electro-optic switch for reconfigurable optical networks-on-chip. Optics Express. 2009;**17**:23979-23808. DOI: 10.1364/OE.17.023793

[18] Reed GT, Mashanovich G, Gardes FY, Thomson DJ. Silicon optical modulators. Nature Photonics. 2010;**4**:518-526. DOI: 10.1038/nphoton.2010.179

[19] Sun DG, Liu P, Hall TJ. Realization for the high electro-optic modulation depth of silicon-on-insulator waveguide devices. In: Proceedings of the IEEE 2013 International Conference on Control Engineering and Information Technology (ICCIT2013); 9-11 August 2013; Nanning, China: IEEE; 2013. pp. 700-703

[20] Chan CC, Tamir T. Angular shift of a Gaussian beam reflected near the Brewster angle. Optics Letters. 1985;**10**:378-380. DOI: 10.1364/OL.10.000378

[21] Tamir T. Nonspecular phenomena in beam fields reflected by multilayered media. Journal of the Optical Society of America. 1986;**A3**:558-565. DOI: 10.1364/JOSAA.3.000558

[22] Aiello A, Merano M, Woerdman JP. Duality between spatial and angular shifts in optical reflection. Physics Review. 2009;**A88**:061801. DOI: 10.1103/PhysRevA.80.061801

[23] Bachmann M, Besse PA, Melchior H. General self-imaging properties in NN multimode interference couplers including phase relations. Applied Optics. 1994;**33**:3905-3911. DOI: 10.1364/AO.33.003905

[24] Soldano LB, Pennings ECM. Optical multimode interference devices based on self-imaging: Principles and applications. Journal of Lightwave Technology. 1995;**13**:615-627. DOI: 10.1109/50.372474

[25] Kawano K, Kitoh T. Schrödinger equation. In: Introduction to Optical Waveguide Analysis—Solving Maxwell's Equations and the Schrödinger Equation. New York, USA: Wiley Interscience; 1995. pp. 615-627. DOI: 10.1002/0471221600.ch7

[26] Himeno A, Terui H, Kobayashi M. Loss measurement and analysis of high-silica reflection bending waveguides. Journal of Lightwave Technology. 1998;**6**:41-46. DOI: 10.1109/50.396

[27] Sun DG, Li X, Wong D, Hu Y, Luo F, Hall TJ. Modeling and numerical analysis for silicon-on-insulator rib waveguide corner. Journal of Lightwave Technology. 2009;**27**:4610-4618. DOI: 10.1109/JLT.2009.2025609